Concept of Operations for Road Weather Connected Vehicle Applications

www.its.dot.gov/index.htm
Draft Final — February 11, 2013
FHWA-JPO-13-047

U.S. Department of Transportation

Produced by FHWA Office of Operations
RITA ITS Joint Program Office
U.S. Department of Transportation

Notice

This document is disseminated under the sponsorship of the U.S. Department of Transportation in the interest of information exchange. The U.S. Government assumes no liability for its contents or use thereof.

The U.S. Government is not endorsing any manufacturers, products, or services cited herein and any trade name that may appear in the work has been included only because it is essential to the contents of the work.

Technical Report Documentation Page

1. Report No. FHWA-JPO-13-047	2. Government Accession No.	3. Recipient's Catalog No.
4. Title and Subtitle Concept of Operations for Road Weather Connected Vehicle Applications		5. Report Date February 11, 2013
		6. Performing Organization Code
7. Author(s) Christopher J. Hill, Ph.D., PMP		8. Performing Organization Report No.
9. Performing Organization Name And Address Booz | Allen | Hamilton 20 M St SE, Suite 900 Washington, DC 20003		10. Work Unit No. (TRAIS)
		11. Contract or Grant No. DTFH61-11-D-00019
12. Sponsoring Agency Name and Address U.S. Department of Transportation Intelligent Transportation Systems – Joint Program Office (ITS JPO) 1200 New Jersey Avenue SE, Washington, DC 20590		13. Type of Report and Period Covered
		14. Sponsoring Agency Code
15. Supplementary Notes Government Task Manager: Paul Pisano, HOTO, Room E86-205, 202-366-1301 Source for all figures and tables in this report is USDOT ITS JPO, and dated February 11, 2013		
16. Abstract Weather has a significant impact on the operations of the nation's roadway system year round. These weather events translate into changes in traffic conditions, roadway safety, travel reliability, operational effectiveness, and productivity. It is, therefore, an important responsibility of traffic managers and maintenance personnel to implement operational strategies that optimize system performance by mitigating the effects of weather on the roadways. Accurate, timely, route-specific weather information allows traffic and maintenance managers to better operate and maintain roads under adverse conditions. Connected vehicle technologies hold the promise to transform road-weather management. Road weather connected vehicle applications will dramatically expand the amount of data that can be used to assess, forecast, and address the impacts that weather has on roads, vehicles, and travelers; fundamentally changing the manner in which weather-sensitive transportation system management and operations are conducted. The US Department of Transportation's Road Weather Management Program has developed this Concept of Operations (ConOps) to define the priorities for connected vehicle-enabled road-weather applications.		
17. Key Words Road Weather, Meteorology, Connected Vehicle, Traffic Management, Traveler Information, Data Collection, Applications	18. Distribution Statement	
19. Security Classif. (of this report) Unclassified	20. Security Classif. (of this page) Unclassified	21. No. of Pages 85	22. Price

Form DOT F 1700.7 (8-72)   Reproduction of completed page authorized

Acknowledgements

The author thanks the USDOT Road Weather Management Program team and participants at the 2012 Road Weather Management Stakeholder Meeting for their valuable input in developing this ConOps.

Table of Contents

Acknowledgements iii
Table of Contents iv
Executive Summary 1
Chapter 1 Scope 2
    Document Security 2
    Identification 2
    Document Overview 2
    Applications Overview 3
Chapter 2 Current Situation 5
    Background and Objectives 5
    Operational Policies and Constraints 6
    Description of Current Situation 7
    Users and Other Involved Personnel 9
        Use of Road Weather Information in Maintenance Operations 9
        Use of Road Weather Information in Traffic Operations 10
        Use of Road Weather Information by Emergency Managers and
        Emergency Responders 12
        Use of Road Weather Information by Motorists and Commercial Vehicle
        Operators 13
Chapter 3 Justification For and Nature of Changes 14
    Justification for Changes 14
        Weather Impacts on Safety 14
        Weather Impacts on Mobility 16
        Weather Impacts on Productivity 16
        The Connected Vehicle Program 17
    Description of Opportunities and Desired Changes 18
        The Connected Vehicle Basic Safety Message 18
        The Vehicle Data Translator 19
        Application of Connected Vehicle Road Weather Data 20
        Road Weather Alerts and Warnings 21
        State and Local Agency-Based Applications 22
        Freight-Based Applications 24
        EMS/First-Responder Applications 25
    Priorities Among Changes 25
    Changes Considered but Not Included 26
Chapter 4 Concepts for the Proposed Applications 27
    Enhanced Maintenance Decision Support System 27
        Description of the Proposed Application 27
        Operational Policies and Constraints 29
        Modes of Operation 30
        User Classes and Other Involved Personnel 30
        Support Environment 31
    Information for Maintenance and Fleet-Management Systems 31

                Description of the Proposed Application 31
                Operational Policies and Constraints 34
                Modes of Operation 34
                User Classes and Other Involved Personnel 34
                Support Environment 35
            Weather-Responsive Traffic-Management Strategies 35
                Description of the Proposed Weather-Responsive VSL Application 35
                Description of the Proposed Weather-Responsive Signalized Intersection
                    Application 37
                Operational Policies and Constraints 39
                Modes of Operation 40
                User Classes and Other Involved Personnel 40
                Support Environment 41
            Motorist Advisories and Warnings 41
                Description of the Proposed Application 41
                Operational Policies and Constraints 44
                Modes of Operation 45
                User Classes and Other Involved Personnel 46
                Support Environment 46
            Information for Freight Carriers 46
                Description of the Proposed Application 46
                Operational Policies and Constraints 49
                Modes of Operation 50
                User Classes and Other Involved Personnel 51
                Support Environment 51
            Information and Routing Support for Emergency Responders 51
                Description of the Proposed Application 51
                Operational Policies and Constraints 54
                Modes of Operation 55
                User Classes and Other Involved Personnel 55
                Support Environment 55

Chapter 5 Operational Scenarios 57
        Scenario for the Enhanced Maintenance Decision Support System 57
            Description 57
            Steps 58
        Scenario for Information for Maintenance and Fleet-Management Systems 59
            Description 59
            Steps 59
        Scenario for Variable Speed Limits for Weather-Responsive Traffic Management 60
            Description 60
            Steps 61
        Scenario for Weather-Responsive Signalized Intersection 62
            Description 62
            Steps 62
        Scenario for Motorist Advisory and Warning System 63
            Description 63
            Steps 64
        Scenario for Information for Freight Carriers 65
            Description 65
            Steps 65
        Scenario for Information and Routing Support System for Emergency
            Responders 66

  Description 66
  Steps 67
Chapter 6 Summary of Impacts 69
  Operational impacts 69
  Organizational Impacts 70
Chapter 7 Analysis of the Proposed System 71
  Summary of Improvements 71
  Disadvantages and Limitations 71
References 73

List of Tables

Table 1: Weather-Related Crash Statistics (Annual Averages) 15

List of Figures

Figure 1: Schematic of the Enhanced-MDSS Application 28
Figure 2: Schematic of Maintenance and Fleet-Management System with
  Connected Vehicle Road Weather Information 33
Figure 3: Schematic of Weather-Responsive VSL System 36
Figure 4: Schematic of Weather-Responsive Signalized Intersection System 38
Figure 5: Schematic of Road Weather Motorist Advisory and Warning System 43
Figure 6: Schematic of Road Weather Advisory and Warning System for Freight
  Carriers 48
Figure 7: Schematic of Road Weather Emergency Responder Dispatching DSS 52
Figure 8: Scenario for the Enhanced Maintenance Decision Support System 58
Figure 9: Scenario for Information for Maintenance and Fleet-Management
  Systems 59
Figure 10: Scenario for Variable Speed Limits for Weather-Responsive Traffic
  Management 61
Figure 11: Scenario for Weather-Responsive Signalized Intersection 62
Figure 12: Scenario for Motorist Advisory and Warning System 64
Figure 13: Scenario for Information for Freight Carriers 65
Figure 14: Scenario for Information and Routing Support System for Emergency
  Responders 67

Executive Summary

Weather has a significant impact on the operations of the nation's roadway system year round. Rain reduces pavement friction. Winter weather can leave pavements snow-covered or icy. Fog, smoke, blowing dust, heavy precipitation, and vehicle spray can restrict visibility. Flooding, snow accumulation, and wind-blown debris can cause lane obstructions.

Weather events may prompt travelers to change departure times, cancel trips, choose an alternate route, or select a different mode. Slick pavements, low visibility, and lane obstructions lead to driving at lower speeds or with increased following distances. These changes in driver behavior can impact the operation of signalized roadways, where traffic signals are timed for clear, dry conditions, through reduced traffic throughputs, increased delays, and increased travel times. Travel reliability for motorists and commercial vehicle operators is affected by a variety of weather conditions. Weather also impacts the operational effectiveness and productivity of traffic management agencies and road maintenance agencies through increased costs and lost time.

It is, therefore, an important responsibility of traffic managers and maintenance personnel to implement operational strategies that optimize system performance by mitigating the effects of weather on the roadways. The operational approaches used by these personnel dictate their needs for weather and road condition information. Accurate, timely, route-specific weather information, allows traffic and maintenance managers to better operate and maintain roads under adverse conditions.

Connected vehicle technologies hold the promise to transform road-weather management. Road weather connected vehicle applications will dramatically expand the amount of data that can be used to assess, forecast, and address the impacts that weather has on roads, vehicles, and travelers; fundamentally changing the manner in which weather-sensitive transportation system management and operations are conducted. The broad availability of road weather data from an immense fleet of mobile sources will vastly improve the ability to detect and forecast road weather and pavement conditions, and will provide the capability to manage road-weather response on specific roadway links.

The U.S. Department of Transportation is providing the national leadership in the connected vehicle program. Connected vehicle research is a multimodal initiative that aims to enable interoperable networked wireless communications among vehicles, the infrastructure, and other wireless devices. Within this program, the USDOT through the Road Weather Management Program has developed this Concept of Operations (ConOps) to define the priorities for connected vehicle-enabled road-weather applications. Six high-priority connected vehicle road weather applications have been identified and described in this document.

U.S. Department of Transportation, Research and Innovative Technology Administration
Intelligent Transportation System Joint Program Office

Chapter 1 Scope

Document Security

This document is not restricted.

Identification

This document is identified as "Concept of Operations for Road Weather Connected Vehicle Applications. Draft Final. February 11, 2013."

Document Overview

The format of this document is consistent with the outline of a concept of operations (CONOPS) document defined in Institute of Electrical and Electronics Engineers Standard 1362-1998. Titles of major sections have, in some instances, been edited to reflect the focus of this document on identifying and describing a set of applications, rather than a single system development effort.

Chapter 2 descr bes the current situation in which road weather information is used to support the management, operations, and maintenance of the roadway network in the United States.

Chapter 3 identifies the need for changes from the current situation and includes descriptions of the impacts of weather events on transportation safety, mobility, and productivity. This section also introduces connected vehicle road weather alerts and warnings applications, state and local agency-based applications, freight-based applications, and Emergency Medical Services (EMS)/first-responder applications.

Chapter 4 provides details of six high-priority connected vehicle road weather applications, while Chapter 5 presents operational scenarios for each application.

Chapter 6 summarizes operational and organizational impacts that may result from the development of the selected applications. Chapter 7 concludes with an analysis of the expected improvements and disadvantages or limitations that may occur following implementation of the applications.

Chapter 1 Scope

Applications Overview

Six high-priority connected vehicle road weather applications are described in this document. The applications can be summarized as follows:

- **Enhanced Maintenance Decision Support System (MDSS).** The MDSS will provide the existing federal prototype MDSS with expanded data acquisition from connected vehicles. Snowplows, agency fleet vehicles, and other vehicles operated by the general public will provide road weather connected vehicle data to the Enhanced-MDSS, which will use this data to generate improved plans and recommendations to maintenance personnel. In turn, enhanced treatment plans and recommendations will be provided to the snowplow operators and drivers of agency maintenance vehicles.

- **Information for Maintenance and Fleet Management Systems.** In this concept, connected vehicle information is more concerned with nonroad weather data. The data collected may include powertrain diagnostic information from maintenance and specialty vehicles, the status of vehicle components, the current location of maintenance vehicles and other equipment, and the types and amounts of materials onboard maintenance vehicles and will be used to automate the inputs to Maintenance and Fleet Management Systems on a year-round basis. In addition, desirable synergies can be achieved if selected data relating to winter maintenance activities, such as the location and status of snowplows or the location and availability of deicing chemicals, can be passed to an Enhanced-MDSS to refine the recommended winter weather response plans and treatment strategies.

- **Weather-Responsive Traffic Management.** Two Weather-Responsive Traffic Management applications are developed. First, connected vehicle systems (CVS) provide opportunities to enhance the operation of variable speed limit (VSL) systems and dramatically improve work zone safety during severe weather events. Additional road weather information can be gathered from connected vehicles and used in algorithms to refine the posted speed limits to reflect prevailing weather and road conditions. Second, CVSs can support the effective operation of signalized intersections when severe weather affects road conditions. Information from connected vehicles can be used to adjust timing intervals in a signal cycle or to select special signal timing plans that are most appropriate for the prevailing conditions.

- **Motorist Advisories and Warnings.** Information on segment-specific weather and road conditions is not broadly available, even though surveys suggest that this information is considered of significant importance to travelers. The ability to gather road weather information from connected vehicles will dramatically change this situation. Information on deteriorating road and weather conditions on specific roadway segments can be pushed to travelers through a variety of means as alerts and advisories within a few minutes. In combination with observations and forecasts from other sources and with additional processing, medium-term advisories of the next 2–12 hours to long-term advisories for more than 12 hours into the future can also be provided to motorists.

- **Information for Freight Carriers.** The ability to gather road weather information from connected vehicles will significantly improve the ability of freight shippers to plan and respond to the impacts of severe weather events and poor road conditions. Information on deteriorating road and weather conditions on specific roadway segments can be pushed to both truck drivers and their dispatchers. In combination with observations and forecasts from other sources and with additional processing, medium- to long-term advisories can also be provided to dispatchers to support routing and scheduling decisions. Because these decisions involve a variety of other factors, such as highway and bridge restrictions, hours-of-service limitations, parking availability, delivery schedules, and—in some instances—the permits the vehicle holds, it is envisioned that the motor carrier firms or their commercial service providers will develop and operate the systems that use the road weather information generated through this concept.

- **Information and Routing Support for Emergency Responders.** Emergency responders, including ambulance operators, paramedics, and fire and rescue organizations, have a compelling need for the short, medium, and long time horizon road weather alerts and warnings. This information can help drivers safely operate their vehicles during severe weather events and under deteriorating road conditions. Emergency responders also have a particular need for information that affects their dispatching and routing decisions. Information on weather-affected travel routes, especially road or lane closures caused by snow, flooding, and wind-blown debris, is particularly important. Low latency road weather information from connected vehicles for specific roadway segments together with information from other surface weather observation systems, such as flooding and high winds, will be used to determine response routes, calculate response times, and influence decisions to hand off an emergency call from one responder to another responder in a different location.

Chapter 2 Current Situation

Background and Objectives

Weather has a significant impact on the operations of the nation's roadway system year round. Rain reduces pavement friction. Winter weather can leave pavements snow covered or icy. Fog, smoke, blowing dust, heavy precipitation, and vehicle spray can restrict visibility. Flooding, snow accumulation, and wind-blown debris can cause lane obstructions. These weather events translate to changes in traffic conditions, roadway safety, travel reliability, operational effectiveness, and productivity.

Traffic conditions may change in a variety of ways. Weather events may prompt travelers to change departure times, cancel trips, choose an alternate route, or select a different mode. Slick pavements, low visibility, and lane obstructions lead to driving at lower speeds or with increased following distances. These changes in driver behavior can affect the operation of signalized roadways, where traffic signals are timed for clear, dry conditions, through reduced traffic throughputs, increased delays, and increased travel times.

Weather affects roadway safety by increasing exposure to hazards and crash risk. Travel reliability for motorists and commercial vehicle operators is affected by a variety of weather conditions. Weather also affects the operational effectiveness and productivity of traffic management agencies and road maintenance agencies through increased costs and lost time.

It is, therefore, an important responsibility of traffic managers and maintenance personnel to implement operational strategies that optimize system performance by mitigating the effects of weather on the roadways. The operational approaches these personnel use dictate their needs for weather and road condition information. Accurate, timely, route-specific weather information allows traffic and maintenance managers to better operate and maintain roads under adverse conditions.

The U.S. Department of Transportation (USDOT) Federal Highway Administration (FHWA) has defined three types of road weather management strategies that can be employed in response to rain, snow, ice, fog, high winds, flooding, tornadoes, hurricanes, and avalanches:

- **Advisory Strategies.** These strategies provide information on prevailing and predicted conditions and impacts to motorists.

- **Control Strategies.** These strategies alter the state of roadway devices to permit or restrict traffic flow and regulate roadway capacity.

- **Treatment Strategies.** These strategies supply resources to roadways to minimize or eliminate weather impacts.

A variety of approaches are available to traffic managers to advise travelers of road weather conditions and weather-related travel restrictions (such as road closures resulting from fog or flooding). Strategies include posting warnings on dynamic message signs (DMS), broadcasting messages via highway advisory radio (HAR), providing road condition reports through interactive traveler information systems such as websites and 511 phone systems, and Public Information Officer interaction with media.

To control traffic flow during adverse weather, traffic managers may regulate lane use (such as lane reversals for evacuations), close hazardous roads and bridges, restrict access on particular roadways to designated vehicle types (e.g., tractor-trailers during high winds), implement variable speed limits, adjust freeway ramp metering rates, or modify traffic signal timings.

Maintenance managers use road weather information and decision support tools to assess the nature and magnitude of winter storms, determine the level of staffing required during a weather event, plan road treatment strategies (e.g., plowing, sanding, chemical applications), and activate anti-icing/deicing systems. Beyond winter weather, maintenance managers are also concerned about the impacts of other events such as sand storms and wildfires that may reduce visibility and create hazardous driving conditions.

Access to high-quality road weather information helps managers improve safety, enhance traffic flow and travel reliability, and increase agency productivity. Weather mitigation strategies enhance roadway safety by reducing crash frequency and severity, restricting access to hazardous roads and encouraging safer driver behavior. Road weather management strategies enhance traffic flow and mobility by allowing the public to make more informed travel decisions, promoting more uniform traffic flow, reducing traffic congestion and delay, and minimizing the time to clear roads of snow and ice. Productivity is increased through better interagency communication and data sharing and by reduced labor, material, and equipment costs for snow and ice control operations.

Operational Policies and Constraints

Operational policies for road weather management activities vary from state to state, both in terms of their detail and their formality. All vary significantly in terms of scope and level of detail; many state transportation agencies have documented policies and procedures that describe strategies for conducting winter and non-winter maintenance activities under various adverse weather conditions. Similar guidelines for the management of traffic operations under adverse weather conditions appear to be less widespread but are gaining ground because of the efforts of the FHWA Weather-Responsive Traffic Management initiative. In many instances, the documented policies and procedures appear to be derived from personnel experience and informal rules of practice. It also appears that documented operational policies are supplemented with undocumented practices.

According to the American Association of State Highway and Transportation Officials, state transportation agencies are increasingly adopting the use of performance-based management approaches. All state departments of transportation (DOT) track asset condition and safety data. The majority of states provide comprehensive performance data to decision makers to both increase accountability to customers and achieve the best possible transportation system performance under current levels of investment. The definition of the performance measures and the formality of reporting again appear to vary from state to state, but weather-related metrics, particularly relating to snow removal during winter storms, are not uncommon.

Chapter 2 Current Situation

Overall, no operational policies related to road weather management are common across the United States. In addition, no policies will specifically constrain the development of connected vehicle road weather applications.

Description of Current Situation

Traffic and maintenance managers use a variety of environmental monitoring systems and other data sources to gather information on weather and related road conditions to make their decisions on how best to mitigate weather impacts. These managers typically use four types of road weather information: atmospheric data (e.g., precipitation type and rate, wind speed and direction), roadway surface data (e.g., surface status and temperature), roadway subsurface data (e.g., subsurface temperature and moisture content), and hydrologic data (e.g., stream levels near roads). These data are generally obtained from various observing system technologies, including fixed sensor stations, transportable sensor stations, mobile sensing devices, and remote sensors.

In addition, traffic and maintenance managers can obtain predictions of environmental conditions from public sources, such as the National Weather Service (NWS), the National Hurricane/Tropical Prediction Center, the National Center for Environmental Prediction, and from private meteorological service providers. Environmental data may also be obtained from mesoscale environmental monitoring networks, or *mesonets,* which integrate and disseminate data from many observing systems (including agricultural, flood monitoring, and aviation networks).

An *environmental sensor station* (ESS) is the field component of an overall Road Weather Information System (RWIS). An ESS comprises one or more sensors measuring atmospheric, surface, subsurface, and water level conditions, while centralized RWIS hardware and software are used to collect and process observation data from numerous ESSs. Traffic and maintenance managers then use environmental observation data from the field to develop route-specific forecasts and provide decision support for various operational actions. State transportation agencies own more than 2,400 ESSs. Most of these stations—more than 2,000—are part of an RWIS used to support winter road maintenance activities. The other stations are deployed for various applications, including traffic management, flood monitoring, and aviation.

Atmospheric data from ESSs include air temperature and humidity, visibility distance, wind speed and direction, precipitation type and rate, and air quality. Roadway surface data include pavement temperature, pavement freeze point, pavement condition (e.g., wet, icy, flooded), pavement chemical concentration, and subsurface conditions (e.g., soil temperature). Water level data include tide levels (e.g., hurricane storm surge); stream, river, and lake levels near roads; and the conditions in areas known to flood during heavy rains or as a result of runoff.

Mobile sensing involves the integration of sensors and other systems onto vehicle platforms. In combination with vehicle location and data communications technologies, mobile sensor systems can be used to sense both pavement conditions (e.g., temperature, friction) and atmospheric conditions (e.g., air temperature). Although less widespread than fixed sensors, several state transportation agencies have deployed maintenance vehicles equipped with mobile environmental sensors. These environmental sensors complement other data collected on vehicles for maintenance purposes, such as snowplow status and material usage. In addition to these efforts by state agencies, a Connected Vehicle Program that could be widely deployed on light and heavy vehicles has the potential to dramatically increase the number of mobile sensor systems across the United States.

Chapter 2 Current Situation

The FHWA Road Weather Management Program (RWMP) is currently demonstrating how weather, road condition, and related vehicle data can be collected, transmitted, processed, and used for decision making through the Integrated Mobile Observations project. In this project, the National Center for Atmospheric Research (NCAR) is partnering with the Minnesota and Nevada DOTs to obtain vehicle data from heavy vehicles, including snowplows, and light-duty vehicles as they carry on routine maintenance functions across their states.

In addition, NWS has sponsored the development of the Mobile Platform Environmental Data (MoPED) system, a mobile sensing system deployed on buses and commercial trucks. Current MoPED data elements comprise road and air temperature, rain intensity, light level, relative humidity, and atmospheric pressure plus derived values of dew point and sea level pressure.

Remote sensors are located at a significant distance from their target. Examples are satellites and radar systems that can be used for surveillance of meteorological conditions. Images and observations from remote sensors are used for weather monitoring and forecasting from local to global scales. Remote sensing is used to quantitatively measure atmospheric temperature and wind patterns, monitor advancing fronts and storms, and image water in all three of its states (i.e., vapor in the air, clouds, snow cover).

Beyond the deployment of the various environmental data-collection systems, initiatives have been undertaken to make the information usable to the transportation community and others. In 2004, USDOT established the *Clarus* Initiative, with its broad goal of reducing the impact of adverse weather conditions on surface transportation users. The *Clarus* Initiative is based on the premise that the integration of a wide variety of weather observing, forecasting, and data management systems combined with robust and continuous data quality checking could serve as the basis for timely, accurate, and reliable weather and road condition information.

A core component of the *Clarus* Initiative is the *Clarus* System. The *Clarus* System is an integrated observation and data-management system that collects near-real-time information from state and local government-owned ESSs together with comprehensive metadata on these systems. The *Clarus* System conducts a variety of quality checks on the data and makes the data available to public and private-sector users and researchers. Currently, 39 states, four local agencies, and five Canadian provinces provide data from more than 2,400 sensors to the *Clarus* System.

The National Oceanic and Atmospheric Administration (NOAA) Meteorological Assimilation Data Ingest System (MADIS) is a similar data-management system that collects data from surface surveillance systems, hydrologic monitoring networks, balloon-borne instruments, Doppler radar, aircraft sensors, and other sources. MADIS leverages partnerships with international agencies; federal, state, and local agencies (including state DOTs); universities; volunteer networks; and the private sector (such as airlines and railroads) to integrate observations from their stations with those of NOAA to provide a finer-density, higher-frequency observational database for use by the meteorological community.

USDOT has also sponsored the development of decision support tools for use by the transportation community. These tools are specifically directed toward the needs of transportation agency users. The goals of the USDOT RWMP in this area acknowledge that decisions affecting the operation and maintenance of the transportation system require decision support tools that directly address the impacts of weather on the roadway system by placing weather and road condition information in a transportation system context.

One such tool is the MDSS, which aids state and local transportation agencies with snow and ice control. MDSS uses weather forecasts, current weather observations, and customized rules of practice to produce road-specific forecasts and recommendations for treatments. Recommendations include a treatment plan (such as plow only, chemical use, or prewetting), recommended chemical application amount, timing of initial and subsequent treatments, and indication of the need to pretreat or post-treat the roads. Today, several companies provide various levels of MDSS capability within their products.

More recently, USDOT has sponsored the development of additional decision support tools that leverage the information contained in the *Clarus* System. Initial development and demonstration of applications included the following:

- A seasonal load-restriction tool that supports state transportation agencies in improving the techniques that lead to the decisions to impose and subsequently lift restrictions on selected roads that are prone to road damage caused by subsurface freezing/thawing processes

- A non-winter maintenance decision support tool that incorporates *Clarus* weather data to assist maintenance-, operations-, and construction-related scheduling decisions during other weather events such as rain, fog, and wind

- A multistate control strategy tool that provides data and strategies to improve coordination among public agencies during adverse weather events, allowing the agencies to proactively respond to situations across jurisdictional boundaries

- Enhanced road weather content for travel advisories using a system that developed appropriate messages regarding the nature, severity, and timing of detected and forecast adverse travel conditions

Initial development work on these tools highlighted the opportunities for expanded system capabilities if additional road weather information can be generated from mobile sensor systems, including connected vehicles.

Users and Other Involved Personnel

Use of Road Weather Information in Maintenance Operations

Maintenance managers obtain and make extensive use of road weather information. This information helps managers make decisions for a variety of winter and non-winter maintenance activities, including decisions about staffing levels, the selection and timing of maintenance activities, and resource management (such as personnel, equipment, and materials) as well as road treatment strategies during winter storms.

Winter road maintenance activities are especially sensitive to weather conditions. During this period of the year, maintenance tasks can often involve snow and ice treatment strategies, including plowing snow, spreading abrasives to improve vehicle traction, and dispensing anti-icing/deicing chemicals to lower the freezing point of precipitation on the pavement. In regions with heavy snowfall, maintenance managers may erect snow fences adjacent to roads to reduce blowing and drifting snow. Another mitigation strategy involves use of slope sensors and avalanche forecasts to minimize landslide and avalanche risks. When a slope becomes unstable because of snow accumulation or soil saturation, roads in the slide path may be closed to allow the controlled release of an avalanche or landslide. After snow, mud and debris are cleared and damaged infrastructure repaired before the affected route can be reopened to traffic.

In mountainous areas during the winter, super-cooled fog can persist in valleys for extended periods. To improve roadway visibility and reduce crash risk, maintenance managers may employ a fog-dispersal strategy. Small amounts of liquid carbon dioxide are sprayed behind maintenance vehicles to encourage precipitation of water droplets in the fog. This strategy includes the application of anti-icing chemicals as fog is dispersed to prevent the precipitation from freezing on road surfaces.

Many non-winter maintenance activities are also affected by weather conditions. Mowing is conducted on a cycle throughout the summer months but will be suspended during heavy rain and thunderstorms. The spraying of herbicides is not conducted during rainstorms or high winds. Striping requires a dry roadway, no high winds, a minimum ambient air temperature, and no immediate likelihood of rain. Surface repairs (such as pothole and seam repairs) using hot mix asphalt need dry pavement with a minimum ambient air temperature and no risk of rain in the short term. Many maintenance activities will also be suspended for lightning storms, tornado forecasts, and periods of low visibility to protect the safety of both maintenance personnel and travelers who may unexpectedly encounter maintenance equipment on or near the roadway.

Use of Road Weather Information in Traffic Operations

The FHWA RWMP is encouraging state and local transportation agencies to be more proactive in the way they manage traffic operations during weather events. Weather-Responsive Traffic Management (WRTM) is the central component of the program's efforts. WRTM involves the implementation of traffic advisory, control, and treatment strategies in direct response to or in anticipation of developing roadway and visibility issues that result from deteriorating or forecast weather conditions.

Over the past 10 years, transportation agencies have implemented various strategies to mitigate the impacts of adverse weather on their operations. These strategies range from simple flashing signs to coordinated traffic-control strategies and regional traveler information. More recently, various new approaches, technologies, and strategies have emerged that hold potential for WRTM, including Active Traffic and Demand Management (ATDM) and Integrated Corridor Management. Operational strategies that traffic managers are currently using include—

- Motorist advisories, alerts, and warnings intended to increase the awareness of the traveler to current and impending weather and pavement conditions. Approaches include active warning systems that warn drivers of unsafe travel conditions through a particular section of roadway, often in remote or isolated locations; pretrip road condition information and forecast systems; and en route weather alerts and pavement condition information

- Speed-management strategies designed to manage speed during inclement weather events. This includes both advisory, which usually involves posting an advisory travel speed deemed safe by the operating agency for the current travel conditions, and regulatory speed-management techniques, which include speed limits that change based on road, traffic, or weather conditions

Chapter 2 Current Situation

- Vehicle restriction strategies involve placing restrictions on the types or characteristics of vehicles using a facility during inclement weather events. These strategies might include size, height, weight, or profile restrictions

- Road restriction strategies restrict the use of a facility during inclement weather to help travelers avoid sections of roadway that are dangerous or would cause substantial delay. Approaches include lane-use restrictions, such as requiring trucks to use a specific lane during inclement weather conditions; parking restrictions, including special parking rules that are implemented during significant snow events that restrict when and where on-street parking is permitted; access control and facility closures; and reversible lane operations, particularly during evacuations

- Traffic signal control strategies involve making modifications or influencing the way traffic signals operate during inclement weather. Approaches in this category include changes to vehicle detector configuration, vehicle clearance intervals, interval and phase duration settings, and implementation of special signal coordination plans designed for inclement weather.

A detailed state-of-the-practice review[1] has identified eight categories of WRTM strategies that comprise over 20 different strategies, as identified here:

- Motorist advisories, alert, and warning systems:

    o Passive warning systems

    o Active warning systems

    o Pretrip condition information and forecast systems

    o En route weather alerts and pavement condition information

- Speed management strategies:

    o Speed advisories

---

[1] U.S. Department of Transportation, Research and Innovative Technology Administration, "Developments in Weather Responsive Traffic Management Strategies," Report No. FHWA-JPO-11-086, June 30, 2011.

- - o   Enforceable speed limits/VSL
- Vehicle restriction strategies:
  - o   Size, height, weight, and profile restrictions
  - o   Tire chains/alternate traction devices
- Road restriction strategies:
  - o   Land use restrictions
  - o   Parking restrictions
  - o   Access control and facility closures
  - o   Contraflow/reversible lane operations
- Traffic signal control strategies:
  - o   Vehicle detector configuration
  - o   Vehicle clearance intervals
  - o   Interval and phase duration settings
  - o   Traffic signal coordination plans
  - o   Ramp control signals/ramp metering
- Traffic incident management:
  - o   Full-function service patrols/courtesy patrols
  - o   Wrecker response contracts
  - o   Quick clearance policies
- Personnel and asset management
- Agency coordination and integration.

Use of Road Weather Information by Emergency Managers and Emergency Responders

Emergency managers, who are responsible the safe movement or evacuation of people during natural or man-made disasters, rely on comprehensive weather and road condition data. Current and predicted weather and road condition information is obtained through RWIS (often through collaboration with transportation agencies or airport operators); water level monitoring systems; and Federal Government sources, such as the National Hurricane/Tropical Prediction Center, commercial weather information providers, and the media. Emergency managers use DSSs that present weather data integrated with population data, topographic data, road and bridge locations, and traffic flow data.

Emergency managers gather weather observations and forecasts to identify hazards and their associated threatened areas and select a response or mitigation strategy. In response to flooding, tornadoes, hurricanes, wild fires, or hazardous material incidents, emergency managers can evacuate vulnerable residents, close threatened roadways and bridges, operate outflow devices to lower water levels, and disseminate information to the public. Many emergency management practices require coordination with traffic managers. Emergency managers may use several control strategies to manage traffic on designated evacuation routes. These strategies include opening shoulder lanes to traffic, contraflow operations to reverse traffic flow in selected freeway lanes, and modified traffic signal timing on arterial routes.

Emergency responders, including fire fighters, ambulance personnel, and paramedics, must routinely operate on roadways affected by adverse weather events. With no option to defer their trips, emergency responders must reach their destinations irrespective of conditions or road closures. Emergency responders rely on routing systems or must make dispatching decisions to hand off an emergency call to another responder, often in the absence of accurate, up-to-date road weather information.

Use of Road Weather Information by Motorists and Commercial Vehicle Operators

Traffic managers disseminate road weather information to road users of all types to influence their travel decisions. Different types of road users have varying information needs. In the event of a road closure, recreational travelers may need alternate route information, while commuters familiar with their route may not. Passenger vehicle drivers are interested in road surface conditions. Commercial vehicle operators who are especially sensitive to time delays and routing may also need information about road restrictions caused by high winds, height and weight limits, or subsurface freeze/thaw conditions. Overall, road weather information allows travelers to make decisions about travel mode, departure time, route selection, vehicle type and equipment, and driving behavior.

Road weather information can be disseminated via roadway infrastructure, telephone systems, websites, and other broadcast media. Roadway systems that are typically controlled by traffic managers use HAR, dynamic message signs, and flashing beacons atop static signs to alert motorists to hazards. Interactive telephone systems and applications on smart phones allow motorists to access road weather information both pretrip and en route. Many state transportation agencies provide general road condition data through toll-free or 511 telephone numbers, websites, and—increasingly—social media.

Chapter 3 Justification For and Nature of Changes

This chapter describes the shortcomings of the existing situation and the opportunities for improvement to the current situation that will motivate development of the new connected vehicle road weather applications.

Justification for Changes

The impacts of weather events on the transportation system have been well analyzed.[2] Adverse weather conditions have been shown to have significant impacts on the safety, mobility, and productivity of transportation system users and roadway operators.

Weather Impacts on Safety

On average, more than 6,301,000 vehicle crashes occur in the United States each year,[3] Twenty-four percent of these crashes, or approximately 1,511,000, are identified as weather related. *Weather-related crashes* are defined as those crashes that occur in adverse weather (such as, rain, sleet, snow, high winds, or fog) or on slick pavement (i.e., wet, snowy/slushy, or icy). On average, 7,130 people are killed and more than 629,000 people are injured in weather-related crashes each year. Although these numbers are showing a downward trend, this data are consistent with overall trends for all traffic fatalities and injuries Table 1 presents an analysis of weather-related crash statistics.

---

[2] http://ops.fhwa.dot.gov/weather/q1_roadimpact.htm. Retrieved March 31, 2012.

[3] Fourteen-year averages from 1995 to 2008 analyzed by Noblis, based on NHTSA data.

Table 1: Weather-Related Crash Statistics (Annual Averages)

Road Weather Conditions	Weather-Related Crash Statistics		
	Annual Rates (Approximately)	*Percentages*	
Wet pavement	1,128,000 crashes	18% of vehicle crashes	75% of weather-related crashes
	507,900 people injured	17% of crash injuries	81% of weather-related crash injuries
	5,500 people killed	13% of crash fatalities	77% of weather-related crash fatalities
Rain	707,000 crashes	11% of vehicle crashes	47% of weather-related crashes
	330,200 people injured	11% of crash injuries	52% of weather-related crash injuries
	3,300 people killed	8% of crash fatalities	46% of weather-related crash fatalities
Snow/sleet	225,000 crashes	4% of vehicle crashes	15% of weather-related crashes
	70,900 people injured	2% of crash injuries	11% of weather-related crash injuries
	870 people killed	2% of crash fatalities	12% of weather-related crash fatalities
Icy pavement	190,100 crashes	3% of vehicle crashes	13% of weather-related crashes
	62,700 people injured	2% of crash injuries	10% of weather-related crash injuries
	680 people killed	2% of crash fatalities	10% of weather-related crash fatalities
Snow/slushy pavement	168,300 crashes	3% of vehicle crashes	11% of weather-related crashes
	47,700 people injured	2% of crash injuries	8% of weather-related crash injuries

U.S. Department of Transportation, Research and Innovative Technology Administration
Intelligent Transportation System Joint Program Office

Road Weather Conditions	Weather-Related Crash Statistics		
	Annual Rates (Approximately)	Percentages	
Fog	620 people killed	1% of crash fatalities	9% of weather-related crash fatalities
	38,000 crashes	1% of vehicle crashes	3% of weather-related crashes
	15,600 people injured	1% of crash injuries	2% of weather-related crash injuries
	600 people killed	1% of crash fatalities	8% of weather-related crash fatalities

Weather Impacts on Mobility

Significant roadway capacity reductions can be caused by flooding or by lane obstruction caused by snow accumulation and wind-blown debris. Road closures and access restrictions resulting from hazardous conditions (such as large trucks in high winds) also decrease roadway capacity.

Weather events can also reduce mobility as well as the effectiveness of traffic signal timing plans on arterials. On signalized arterial routes, speed reductions can range from 10 percent to 25 percent on wet pavement and from 30 percent to 40 percent on snow-covered or slushy pavement. Average arterial traffic volumes can decrease by 15 percent to 30 percent depending on road weather conditions and time of day. Travel reliability is significantly affected by the impacts of weather events on the roadway. For example, travel time delay on arterials can increase by 11 percent to 50 percent, and start-up delay can increase by 5 percent to 50 percent depending on the severity of the weather impact.

On freeways, light rain or snow can reduce average speed by 3 percent to 13 percent, while heavy rain can decrease average speed by 3 percent to 16 percent; in heavy snow, average freeway speeds can decline by 5 percent to 40 percent. Low visibility can cause speed reductions of 10 percent to 12 percent. Freeway capacity reductions can also be significant— 4–11 percent in light rain, 10–30 percent in heavy rain, 12–27 percent in heavy snow, and by 12 percent in low visibility.

Overall, it has been estimated that 23 percent of the nonrecurrent delay on highways across the nation is because of the impacts associated with snow, ice, and fog. This amounts to an estimated 544 million vehicle-hours of delay per year.

Weather Impacts on Productivity

Adverse weather can also increase the operating and maintenance costs of road maintenance agencies, traffic management agencies, emergency management agencies, law enforcement agencies, and commercial vehicle operators. Winter road maintenance activities account for roughly

20 percent of state transportation agency maintenance budgets. Each year, state and local agencies spend more than $2.3 billion on snow and ice control operations.

Each year, trucking companies lose an estimated 32.6 billion vehicle-hours because of weather-related congestion in the nation's top 281 metropolitan areas. The estimated cost of weather-related delay to trucking companies is $3.1 billion annually in the nation's 50 largest cities.

The availability of accurate, up-to-date road weather observations that are tailored to the needs of roadway operators together with the decision support tools that place the observation data in a transportation system operations and management context can play a significant role in helping better prepare roadway operators and users of the transportation system for adverse weather conditions. In turn, this approach has the potential to improve safety, mobility, and productivity. The FHWA RWMP has already undertaken significant work to acquire, quality check, and make available road weather observations from fixed, mobile, and remote sensing systems.

The Connected Vehicle Program

USDOT is also providing the national leadership in the connected vehicle program. Connected vehicle research is a multimodal initiative that aims to enable interoperable networked wireless communications among vehicles, the infrastructure, and other wireless devices. Connected vehicle applications will provide connectivity—

- Among vehicles to enable crash prevention

- Between vehicles and the infrastructure to enable safety, mobility, and environmental benefits

- Among vehicles, infrastructure, and wireless devices to provide continuous real-time connectivity to all system users.

Connected vehicle safety applications are intended to increase situational awareness and reduce or eliminate crashes through vehicle-to-vehicle (V2V) and vehicle-to-infrastructure (V2I) data transmission that will support driver advisories, driver warnings, and vehicle and infrastructure controls. These technologies may potentially address up to 82 percent of crash scenarios with unimpaired drivers, preventing tens of thousands of automobile crashes every year.

Connected vehicle mobility applications will provide a connected, data-rich travel environment. The connected vehicle network will capture **real-time data** from equipment located on-board vehicles and within the infrastructure. The data are transmitted wirelessly and are used by transportation managers in a wide range of applications to manage the transportation system for optimum performance.

Connected vehicle environmental applications will both generate and capture environmentally relevant real-time transportation data and use these data to create actionable information to facilitate "green" transportation choices. For instance, informed travelers may decide to avoid congested routes, take alternate routes, take public transit, or reschedule their trip. Data generated from connected vehicle systems can also provide operators with detailed, real-time information that can be used to improve system operations. On-board equipment may also advise vehicle owners on how to optimize the operation and maintenance of their vehicle for maximum fuel efficiency.

Connected vehicle technologies hold the promise to transform road weather management. Road weather connected vehicle applications will dramatically expand the amount of data that can be used to assess, forecast, and address the impacts that weather has on roads, vehicles, and travelers, fundamentally changing the manner in which weather-sensitive transportation system management and operations are conducted. The broad availability of road weather data from an immense fleet of mobile sources will vastly improve the ability to detect and forecast road weather and pavement conditions and will provide the capability to manage road weather response on specific roadway links.

Road weather connected vehicle applications are uniquely cross-cutting, affecting the research efforts in many other connected vehicle program areas. The applications developed in the road weather connected vehicle area will also capitalize on the current *Clarus* research, building from an existing integrated network of road weather information to create a data environment comprising both fixed and mobile sources that will be invaluable for research and operational activities. The road weather connected vehicle applications will also leverage and enhance the FHWA RWMP investments that have been made in strategies and tools for traffic and maintenance management. In this way, road weather information from connected vehicle sources will enable exciting new capabilities in the maintenance decision support tools and weather-responsive traffic management strategies that are already implemented and proven.

It is difficult to imagine any transportation system activity that is unaffected by high-impact weather events and the resulting deterioration of road conditions. Connected vehicle road weather data can therefore be viewed as an additional source of information beyond traditional weather products and services that will be of unprecedented importance to other connected vehicle applications in all areas of safety, mobility, and environmental improvement. Applications such as intersection collision avoidance, signalized intersection control, speed warnings, and traveler information dissemination can all be significantly enhanced when they take current road weather information into account.

The availability of connected vehicle road weather information will also create new opportunities to provide applications to other stakeholders, including those outside the traditional surface transportation community, for whom the condition and performance of the roadway system is particularly important. Freight shippers, public safety agencies, and EMS will be among the significant beneficiaries of road weather information from connected vehicle sources. Information on current weather and pavement conditions on specific roadway links will open transformative applications for these constituents in areas of routing, scheduling, and response capabilities.

Description of Opportunities and Desired Changes

Efforts have already been undertaken to define the types of data that can be acquired from CVSs and that will support the development of road weather-related applications. Two particular activities are described here: the definition of a connected vehicle basic safety message (BSM) and the development of a vehicle data translator (VDT) within the RWMP.

The Connected Vehicle Basic Safety Message

Connected vehicle V2V safety applications heavily rely on the BSM, which is one of the messages defined in the Society of Automotive Engineers (SAE) Standard J2735, *Dedicated Short Range Communications (DSRC) Message Set Dictionary* (November 2009). The development of the BSM is ongoing and evolving. At the time of writing, the BSM consists of two parts, with the following characteristics:

- BSM Part 1 contains core data elements, including vehicle position, heading, speed, acceleration, steering wheel angle, and vehicle size. It is transmitted at a rate of about 10 times per second.

- BSM Part 2 contains a variable set of data elements drawn from an extensive list of optional elements. They are selected based on event triggers (such as when the antilock braking system [ABS] is activated). BSM Part 2 data elements are added to Part 1 and sent as part of the BSM message but are transmitted less frequently to conserve data communications bandwidth.

It is important to note that even if a data element is defined in BSM Part 2 of the SAE J2735 standard, it does not necessarily mean that vehicle manufacturers will provide it. Most of the Part 2 data elements are defined as optional information in the standard. Some of the Part 2 data elements are currently available on the internal data bus of some vehicles; others are not.

Appendix B contains a table listing data elements found in BSM Part 1 or Part 2. The table also contains the results of an analysis[4] that identifies whether the element may be useful in determining road weather conditions. This shows that most desired Part 2 elements are weather related.

It should be noted that USDOT has requested that certain weather data be incorporated into the SAE J2735 SE message set. The requested data elements, their ranges, and the resolution of the request are presented in Appendix C.

The Vehicle Data Translator

The development of a VDT has been undertaken through the FHWA RWMP. The VDT is a system that ingests and processes mobile data available on the vehicle and combines them with ancillary weather data sources. The earliest versions of the VDT were developed using nine to 11 vehicles operating in a development test environment in Detroit during the winter and spring of 2009 and 2010. Development and validation of VDT Version 3.0 is underway at the time of writing.

A long-term view of the connected vehicle program includes the collection of data by millions of passenger and commercial vehicles. However, for this data to be useful to the broad community of

---

[4] USDOT RITA ITS Joint Program Office, "Vehicle information exchange needs for mobility applications," FHWA-JPO-12-021, February 13, 2012.

stakeholders, it must be acquired, and then processed into meaningful, actionable information. The VDT inputs two types of data:

- *Mobile data* are all data originating from a vehicle, whether native to the Controller Access Network Bus (CANBus) or as an add-on sensor (e.g., pavement temperature sensor mounted to a vehicle).

- *Ancillary data* represent all other data, such as surface weather stations, model output, satellite data, and radar data.

Current development efforts indicate that the VDT will function best where a minimum set of data elements is available. These comprise environmental and vehicle status data elements from the mobile source, including external air temperature, wiper status, headlight status, ABS and traction control system status, rate of change of steering wheel, vehicle velocity, date, time, location, vehicle heading, and pavement temperature plus ancillary data elements of radar, satellite, and surface station data from fixed data sources.

When the VDT has acquired data, they undergo quality checking followed by the application of various algorithms to create useful road weather information. Algorithms in development through VDT Version 3.0 include—

- A precipitation algorithm that will provide an assessment of the type and intensity (amount/hour) or accumulation rate of precipitation that is falling to the road surface by road segment. It is anticipated that the algorithm will identify four precipitation types—rain, snow, ice/mixed, and hail—and will distinguish between light/moderate and heavy rates of each precipitation type

- A pavement condition algorithm is being developed to derive the pavement condition on a segment of roadway from the vehicle observations. Pavement conditions being considered are dry, wet, road splash, snow, icy/slick, and hydroplaning risk

- A visibility algorithm is being designed to provide additional information by road segment on both a general decrease in visibility and more specific visibility issues. This approach is intended to report visibility as normal or low and potentially identify specific hazards, including dense fog, heavy rain, blowing snow, and smoke.

Application of Connected Vehicle Road Weather Data

The emergence of new sources of road weather information from connected vehicles opens opportunities to dramatically enhance existing systems and to create transformative new applications for the data. In general, three broad categories of opportunity have been identified:

- Use of connected vehicle data to enhance existing strategies, tools, and systems that are focused on the needs of the traffic and maintenance management community to respond to the impacts of adverse weather on the roadways

- Use of connected vehicle data to create new strategies, tools, and systems that are focused on the road weather information needs of other stakeholders

- Use of connected vehicle road weather information to bring additional capabilities to other connected vehicle safety, mobility, and environmental applications. Within the connected vehicle research program at USDOT, additional definition and development work is underway on a variety of safety and dynamic mobility applications. In a number of instances, these activities acknowledge that road weather information might contribute to the effectiveness of the application. It is, therefore, suggested that connected vehicle road weather information, with its potentially dense, roadway segment-specific nature and short time horizon, would especially enhance these applications and should be considered a key component of their development.

Cross-cutting these categories, the following taxonomy of application areas has been developed:

- Road weather alerts and warnings
- State and local agency-based applications
- Freight-based applications
- EMS/first-responder applications.

This section of the CONOPS introduces and briefly describes the applications that have been identified in each of these application areas. High-priority applications are selected in the following section, and are then explored in greater detail in Chapter 5.

Road Weather Alerts and Warnings

Motorist Advisories and Warnings

Although motorists now have access to multiple sources of travel information through roadway infrastructure, radio broadcasts, phone systems, and applications on their personal mobile devices, a recent analysis[5] suggests that this community is underserved with road weather information. The analysis further indicates that weather information for roadways of interest is an especially high priority for these users—more important than other forms of travel information, including incident and travel time reports.

A road weather connected vehicle application would push roadway link-specific information to users' in-vehicle equipment or personal wireless devices. As a minimum, users would receive road weather alerts and warnings within a short time horizon of adverse conditions being detected by mobile data

---

[5] American Meteorological Society, *Realizing the Potential of Vehicle-Based Observations*, 2011.

sources. These conditions may include precipitation types and rates, road surface slickness, and low visibility.

Real-time mobile source data would also be combined and processed with forecast information and data from other fixed and remote sensors to provide medium to longer-term alerts and warnings to users. Opportunities exist with this application for commercial service providers to use these road weather alerts and warnings through various onboard or off-board processing capabilities to deliver routing and other traveler information services to subscribers.

Enable Advanced Traveler Information Systems

The Enable Advanced Traveler Information Systems (EnableATIS) represents a bundle of connected vehicle applications currently being developed within the USDOT Dynamic Mobility Applications (DMA) program. This bundle of applications seeks to provide a framework for multisource, multimodal data to enable the development of new advanced traveler information applications and strategies. EnableATIS envisions a traveler information services framework with a pool of real-time data through connected vehicles, public and private systems, and user-generated content. Current work on EnableATIS is not defining specific applications but is instead formalizing the framework to support diverse traveler information solutions. The existing work recognizes the importance of road weather information as a component of the real-time data pool. However, the current EnableATIS work does not emphasize the importance of road weather information at the level of timeliness, accuracy, and relevance that drivers request and that will be facilitated through focused road weather connected vehicle applications.

State and Local Agency-Based Applications

Enhanced Maintenance Decision Support System

MDSS is an existing decision support tool, described earlier in this report, that maintenance managers use to develop treatment and response plans to winter storms and other winter weather events. Available MDSS solutions typically acquire data from fixed and remote sensors for use in various weather and pavement temperature models. Although the federal MDSS prototype acquires data from mobile sources, it is exclusively automatic vehicle location data used to display current snowplow locations.

Mobile sensor data, both from the general vehicle fleet and from additional, specialized sensors on plows and other maintenance vehicles, can be used to expand the capabilities of MDSS. In particular, a denser, more comprehensive set of mobile observations will provide data for more accurate model runs and forecasts in complex terrain or in areas where sensor networks are particularly sparse. Data from specialized sensors on agency-controlled vehicles, such as real-time measures of salinity and freeze point on specific segments of roadway, will help MDSS optimize and communicate treatment strategies to plow operators.

Information for Maintenance and Fleet Management Systems

Maintenance and fleet management systems are typically software-based systems used to manage an agency's vehicle and material assets. These systems will monitor the status (e.g., vehicles in or out of service), locations, quantities, and usage of assets, as appropriate. The information is used for scheduling maintenance of vehicles, ordering materials, and deploying assets to the required

Chapter 3 Justification for and Nature of Changes

locations. Certain nonweather-related connected vehicle data, such as maintenance diagnostics or vehicle location, could provide valuable real-time information to these systems.

In addition, integration between maintenance management systems and MDSS may provide some opportunities, especially when using real-time connected vehicle data. Vehicle status and location information can be used in MDSS to support the development of treatment plans, while outputs of treatment strategies, including the types and quantity of materials used, can be used to update information in maintenance management systems.

Weather-Responsive Traffic Management

Several weather-responsive traffic-management strategies were identified earlier in this document and are already well established. Several of these could benefit from the additional sources of mobile road weather data afforded by connected vehicles. The denser and more comprehensive network of observations will assist traffic managers in their decisions to implement restrictions.

Weather-responsive traffic-management strategies that include pretrip and en route advisories and warnings will be enhanced through the availability of connected vehicle road weather data. These applications focus on the needs of motorists and are described elsewhere.

Another category of weather-responsive traffic-management strategies involves signalized intersection controls. There is extensive work on this topic in other areas of the connected vehicle program that is discussed in a subsequent section of this chapter. However, this area is ripe for additional exploration of the impacts of road weather-specific connected vehicle data and is discussed in greater detail later in this document.

The use of connected vehicle road weather information in certain ATDM applications is not addressed in other connected vehicle program areas and will be discussed in this section. In particular, connected vehicle road weather information would facilitate the development of weather-responsive speed advisories or VSLs. With appropriate state or local legislation, VSLs are enforceable by public safety agencies.

A VSL application would use data acquire from connected vehicle mobile sources to determine precipitation types and amounts, visibility, or road surface slickness for segments of the roadway network under VSL control. These data would be combined with other information on prevailing traffic volumes and speeds. Algorithms would be developed to determine appropriate travel speeds under the current traffic, weather, and road conditions. Speed limit information would be displayed on suitable roadway infrastructure and potentially could be provided directly to drivers of suitably equipped vehicles in the form of alerts or in-vehicle signage.

Intelligent Network Flow Optimization

The Intelligent Network Flow Optimization (INFLO) bundle of applications is under development within the DMA program area. This bundle consists of three applications: speed harmonization, queue warning, and cooperative adaptive cruise control. The objective of speed harmonization is to dynamically adjust and coordinate maximum appropriate vehicle speeds in response to downstream congestion, incidents, and weather or road conditions to maximize traffic throughput and reduce crashes. The objective of queue warning is to provide a vehicle operator sufficient warning of impending queue backup to brake safely, change lanes, or modify route such that secondary collisions can be minimized. A queue backup can occur as a result of several conditions, including adverse weather. The objective of cooperative adaptive cruise control is to dynamically and automatically coordinate cruise control speeds among platooning vehicles to significantly increase traffic throughput. The need for road weather information from both fixed and mobile sources is acknowledged in the current concept development work for INFLO.

Traffic Signal and Stop Sign Violation Warnings

Within the connected vehicle V2I safety application program, traffic signal and stop sign violation warning applications are under development that are intended to predict whether a driver approaching a signal or stop sign will be in violation, and then issue a warning to the driver. Adverse weather conditions (such as precipitation or a slick road surface) would affect when the warning should be issued. Connected vehicle road weather information could be integrated into the warning algorithm.

Curve Speed Warnings and Rollover Warnings

Also within the V2I safety application program, curve speed warning and rollover warning applications are being developed that will aid drivers in negotiating curves at appropriate speeds. Icy roads or high winds would influence the timing of the warning and the recommended speed. Road weather information from connected vehicles could support the development of this application.

Freight-Based Applications

Information for Freight Shippers

The needs of freight shippers for road weather information are naturally different from other motorists. In certain instances, roadway restrictions, such as lane closures and seasonal road closures, apply uniquely to motor carriers. In other situations, certain adverse weather events, such has high winds, have a far greater impact on high-profile trucks than they do on passenger vehicles. Commercial freight shippers are also especially sensitive to travel delays, which translate directly into productivity losses and higher costs and have a particular impact on just-in-time deliveries. Road closures also have special impacts where rerouting may not always be possible because of bridge height or highway weight restrictions or restrictions placed on special haulers such as hazardous material carriers.

The Freight Advanced Traveler Information System (FRATIS) project discusses these information needs together with the application of connected vehicle technologies to optimize intermodal drayage operations and acknowledges the importance of road weather information as a component of the freight industry's needs. However, the current FRATIS work does not emphasize the importance of road weather information at the level of timeliness, accuracy, and relevance that commercial vehicle operators are requesting and that will be facilitated through focused road weather connected vehicle applications. Road weather connected vehicle freight applications must therefore ensure that the identified needs in FRATIS are met but also identify and support the deeper and broader road weather information needs of the freight community.

A road weather connected vehicle application for freight shippers must accommodate the different information needs of the driver and the dispatcher. Short-term alerts on precipitation type and rate; road surface slickness; high winds; low visibility; and the presence of thunderstorms, hail, and tornadoes are all of immediate concern to drivers, who must operate their vehicles safely under

Chapter 3 Justification for and Nature of Changes

deteriorating conditions and make decisions about their hours of service. This information, together with longer-term regional forecasts and information about road closures and restrictions, is important to fleet managers and dispatchers, who will use it to make decisions about schedule and routing changes.

Opportunities in this application area exist for commercial service providers to integrate road weather data from connected vehicles with other fleet-management and decision support tools that motor carrier companies use.

EMS/First-Responder Applications

Information and Routing Support for Emergency Responders

Emergency responders, including fire and rescue organizations, paramedics, and ambulance operators, represent a unique community of stakeholders for connected vehicle road weather information. Unlike many other users of the transportation system, this group of constituents has no opportunity to cancel or defer trips. In fact, this group of users is particularly called upon to use the roadways during adverse weather situations. To further exacerbate the challenges for this user group, the lowest volumes of mobile sensor observations will inevitably be generated when other drivers stay off the roadway during severe weather events.

It is important for first responders to have access to road condition information through short-term alerts and warnings in a manner similar to other users. However, situational awareness is especially important to emergency vehicle drivers who cannot avoid roadways affected by severe weather events but need to know what they are getting into so that drivers can safely operate their vehicles. Emergency responders are also especially sensitive to information that affects routing and dispatching decisions. In particular, lane and road obstructions caused by snow or wind-blown debris, for example, is important information used to determine routes, calculate response times, and support decisions to pass a call from one emergency responder to another. Overall, this group of users demands road weather information at a level of timeliness, accuracy, and relevance that traditional weather products and services cannot provide.

A particular challenge in this application will be the identification of the effects of adverse weather events that cannot be directly measured with mobile, fixed, or remote weather sensors. These effects may include downed trees and power lines that cause road closures. Nonweather-related connected vehicle data that can identify vehicle stops and queue build-ups will be necessary in this situation.

Priorities Among Changes

The high-priority areas for further development of connected vehicle road weather applications appear to be those that are not being pursued in other parts of the connected vehicle program. The applications that will be explored in greater detail in this CONOPS are—

- Motorist advisories and warnings

- Enhanced maintenance DSS

- Information for maintenance and fleet-management systems

- Weather-responsive traffic-management strategies, including VSLs and signalized intersection control

- Information for freight shippers

- Information and routing support for emergency responders.

Changes Considered but Not Included

Safety and mobility applications being developed elsewhere in the connected vehicle program were considered and have been discussed earlier in this document. It appears that the relevance of road weather information is being considered in these development efforts, although not necessarily at the level of timeliness, accuracy, and relevance that can be provided through road weather connected vehicle applications. Therefore, although the specific DMA applications of EnableATIS, INFLO, and FRATIS are not included for further discussion in this document, the broader road weather information needs of the constituents of these applications (such as road weather advisories for freight operators) are explored in greater detail in the next chapter of this document.

Chapter 4 Concepts for the Proposed Applications

This chapter provides a discussion of each high-priority application in terms of scope, users, goals and objectives, capabilities, various operational modes, how and where it will operate, what it will interface with, and their lines of communication.

Enhanced Maintenance Decision Support System

Description of the Proposed Application

Previous chapters of this report have descr bed the existing MDSS. State and local transportation agencies must handle multiple tasks and process large amounts of information in the development of their response and treatment plans during winter weather conditions and events. MDSS is a decision support tool that integrates relevant roadway segment-specific road weather forecasts, coded rules of practice, and maintenance resource data to provide winter maintenance managers with recommended road treatment strategies. Use of MDSS allows maintenance personnel to make both strategic and tactical decisions that improve roadway levels of service and safety and are more efficient in the use of labor, equipment, and chemicals. The discussion in this chapter relates to potential enhancements to the federal prototype MDSS. It does not mean that these ideas and approaches have not been identified or are being considered by private-sector providers of MDSS solutions.

In the existing system, meteorological and road observations together with output from national weather prediction models provide the input data to MDSS. The input data are then used in a Road Weather Forecast System to generate a forecast of weather impacts on the roadways. Elements of this forecast include air temperatures; wind speeds; relative humidity and dew point; and the types, intensity, and amount of precipitation. The forecast weather impacts are used in a Road Condition and Treatment Module to predict specific roadway conditions, including pavement temperature and pavement condition (such as measures of snow depth and pavement friction), bridge frost potential, and blowing snow potential. This module also generates recommended treatment plans, such as plans for plowing, chemical use, and prewetting; recommendations on the amount of chemicals to be applied; and timing of initial and subsequent treatments.

The effectiveness of the MDSS recommendations is directly related to the quality and extent of the input data. Weather forecast models are effective in providing information on large-scale events but may be less effective in providing accurate forecasts for short-lived or small-scale events and do not provide tailored information on the impacts of weather events on the roadways. Agencies supplement the information from forecast models with observation data from other fixed and remote sensors, including their own networks of environmental sensor stations and from Federal Aviation Administration surface weather observation stations. However, these observation stations are often sparsely deployed and do not provide a detailed representation of surface weather conditions in complex terrain.

The Enhanced-MDSS application concept will change this situation by providing the system with expanded data acquisition from connected vehicles. Snowplows, agency fleet vehicles, and other vehicles operated by the general public will provide road weather connected vehicle data to the Enhanced-MDSS, which will use this data to generate improved plans and recommendations to maintenance personnel. In turn, enhanced treatment plans and recommendations will be provided back to the snowplow operators and drivers of other agency maintenance vehicles. In addition, connected vehicles will continue to gather road condition information after treatment plans have been implemented. These vehicles will provide a continuous measure of the level of service of a roadway and the outcome of the initial treatment. Figure 1 provides a schematic of how the Enhanced-MDSS application could work.

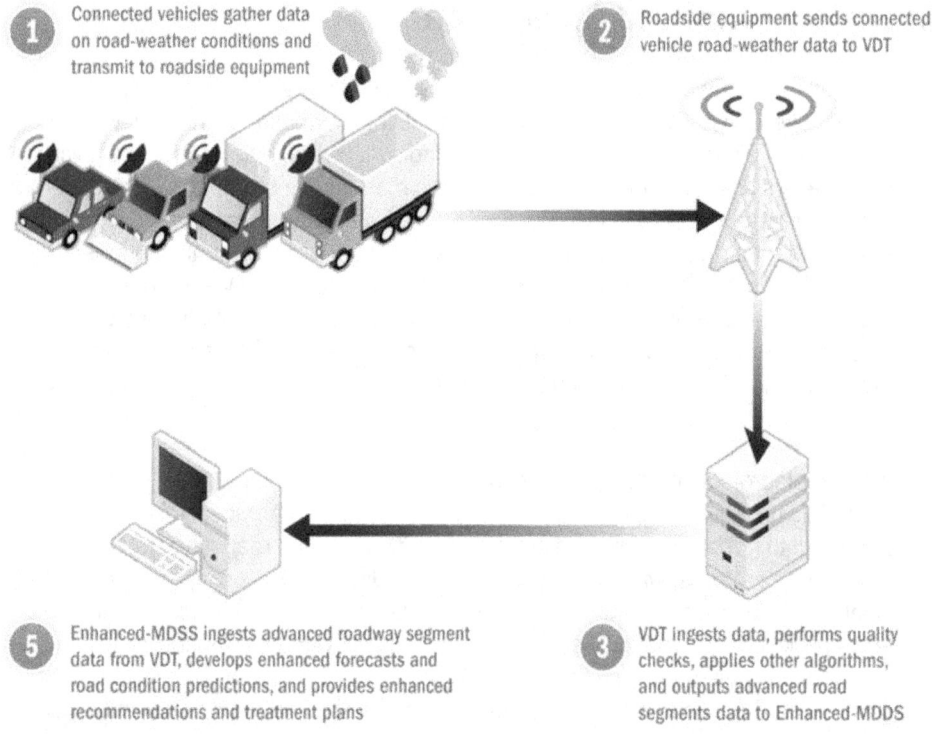

Figure 1: Schematic of the Enhanced-MDSS Application

This application will consist of a series of subsystems:

Data Acquisition Subsystem

This subsystem is made up of the connected vehicles, with their associated onboard equipment and the necessary roadside infrastructure. Two classes of connected vehicle are anticipated for the application: (1) vehicles operated by the general public and commercial entities (including passenger cars and trucks) and (2) specialty vehicles and public fleet vehicles (such as snowplows, maintenance trucks, and other agency pool vehicles). It is assumed that passenger cars and commercial trucks will provide data elements specified in BSM Parts 1 and 2 (including the weather-related data elements in BSM Part 2), while agency-controlled vehicles will provide these data elements and, optionally, additional data elements from specialty sensors installed on selected vehicles (e.g., sensors to measure salinity at the roadway surface).

Data Processing Subsystem

Connected vehicle road weather data will be communicated via data backhaul to a VDT. The VDT will process the raw data and generate road segment-based data outputs that will be available to the Enhanced-MDSS. The VDT data outputs will be assimilated in MDSS back-end processors for use in the various weather and pavement temperature models and will be available to MDSS users. Outputs from the VDT will include—

- Weather variables, such as air temperature, barometric pressure, and dew point

- Road weather variables, such as pavement temperature, friction, and salinity

- Other variables, such as average vehicle speeds, ABS activation events, and vehicle stability and traction control activation events

- Inferred variables (from VDT algorithms), such as slickness, visibility, and precipitation rate and type.

VDT output data will supplement other data sources (such as data from national weather models and from surface weather and road weather observation systems used in the existing MDSS) and will be used in an enhanced Road Weather Forecast System and Road Condition and Treatment Module to predict roadway conditions and provide recommended treatment plans.

User Interface System

State and local agency maintenance personnel will interact with the Enhanced-MDSS in a manner similar to the existing MDSS. Additional decision support tools may be developed that use the detailed roadway segment-specific data that is newly available from the connected vehicle data sources and the VDT. New techniques will be developed to communicate the enhanced treatment plans and recommendations back to snowplow and maintenance fleet vehicle operators.

Operational Policies and Constraints

This section identifies current constraints and potential future constraints and risks.

Data Availability

The effectiveness of this application is predicated on the availability of connected vehicle road weather information. This assumes a broad penetration of connected vehicle onboard equipment into the national vehicle fleet and the availability of an appropriate roadside and data backhaul infrastructure. It further assumes the willingness of state and local agencies to deploy connected vehicle devices and potentially other specialty sensors into the vehicles under their control. Additional research will clarify the levels of agency and general fleet penetration required to generate sufficient data for the application to work effectively.

In addition, in this application, data is especially desired in advance of predicted winter storms and during other severe winter weather events. During these periods, drivers may be encouraged and inclined to avoid travel, which may affect the availability of the required road weather data.

VDT Implementation

This application will require participating agencies or their contractors to implement and operate the VDT. The VDT is currently in a development phase; therefore, the impacts of this requirement are currently unknown.

MDSS Enhancements

The existing federal prototype MDSS will need to undergo enhancements to be able to assimilate and use the connected vehicle road weather information in its operation. This will require additional research and development that must be defined and performed.

Deployment Coverage

A sufficiently dense network of roadside equipment with adequate geographic coverage will be required to gather connected vehicle road weather data that is effective for the concept. This will be particularly important in areas of complex terrain or where information on short roadway segments is desired.

Modes of Operation

The typical modes of operation for the Enhanced-MDSS concept are—

Normal Mode. In the normal operating mode, the Enhanced-MDSS will be available in advance of and during all winter weather events, with all designed functionality available.

Degraded Mode. In this mode, some functions are not working properly or may not be available. This could result from many different situations. In the event that connected vehicle road weather data are not available or the VDT is not functioning, the system will operate in the manner of the existing MDSS.

Maintenance Mode. During system maintenance, some subsystems and their functionality may not be available. This mode is similar to Degraded Mode, except that during Maintenance Mode it may be possible to bring the subsystems back into operation, if needed.

User Classes and Other Involved Personnel

Vehicle Operators. In most cases, vehicle operators will be passive participants in the operation of the Enhanced-MDSS. While operating their vehicles, onboard equipment will collect connected vehicle road weather information and communicate this information to appropriate roadside equipment. Most vehicle operators will not be recipients of the information that the Enhanced-MDSS generates.

Snowplow and Maintenance Vehicle Operators. In this special class of vehicle operators, the drivers will be passive participants in the collection and communication of connected vehicle road weather data. However, these vehicle operators will be intended recipients of the information that the Enhanced-MDSS generates. Operators of these vehicles will interact with appropriate in-vehicle devices to receive instructions on their actions during winter weather events.

Maintenance Personnel. This group of users will interact with the Enhanced-MDSS. They will receive recommendations on winter weather treatment strategies from the Enhanced-MDSS based on roadway segment-specific information from the VDT. They will use the decision support tools available through the Enhanced-MDSS and direct the actions of the snowplow and other maintenance vehicle operators based on the system outputs.

Support Environment

The Enhanced-MDSS concept will operate within the overall CVS. As such, the Enhanced-MDSS requires the deployment of connected vehicle onboard equipment and a DSRC roadside infrastructure or other wireless communications system, such as cellular; access to the certificate management entities defined for the CVS; and suitable data communications backhaul.

It is assumed that the systems operating the VDT will be deployed coincident with the data processing and communications systems required to operate MDSS within a state or local government facility or could be operated by a value-added service provider or other contractor. Appropriate systems administrators, system maintenance, and IT personnel will be required.

A suitable communications infrastructure, in common with the existing MDSS approach, will be required for the maintenance personnel using the Enhanced-MDSS to provide actions and directions to snowplow operators and other maintenance vehicle operators.

Information for Maintenance and Fleet-Management Systems

Description of the Proposed Application

This concept is viewed as both a stand-alone application and as an adjunct to the Enhanced-MDSS application described in the previous section. Maintenance and fleet-management systems are primarily concerned with the control of a transportation agency's physical assets, including its maintenance vehicles and materials (which can include the materials for roadway and pavement repair, chemicals and supplies for roadside vegetation control, and the chemicals and related materials used to control roadway icing and snow removal). These systems are used for a variety of purposes, including the management of material and fuel usage and purchases; managing the allocation of staff and other resources; equipment maintenance planning and scheduling; budget monitoring and forecasting; and long-term acquisition support and procurements, including life-cycle cost analyses for equipment and vehicles.

In this concept, connected vehicle information is more concerned with non-road weather data. The data collected may include powertrain diagnostic information from maintenance and specialty vehicles, the status of vehicle components such as plow blades and spreaders, the current location of maintenance vehicles and other equipment, and the types and amounts of materials onboard maintenance vehicles. In addition, connected vehicle information can be used to automate end-of-route and end-of-shift reports, where information on mileage and fuel usage is important. These types of information are key to automating the inputs to Maintenance and Fleet-Management Systems on a year-round basis.

In addition, desirable synergies can be achieved if selected data relating to winter maintenance activities, such as the location and status of snowplows or the location and availability of deicing chemicals, can be passed to an Enhanced-MDSS to refine the recommended winter weather response plans and treatment strategies. Real-time information from mobile assets on material usage can be particularly important to tactical decision making during winter storms. Information from CVSs can also assist the decision-making process as it relates to the selection of the appropriate equipment for the current conditions and situations on the roadways and in decisions on how to move equipment from one geographic location to another to respond to needs. Finally, information from connected vehicles can be used to determine whether drivers and operators are correctly following treatment plans in terms of routes cleared and appropriate chemical distribution.

Figure 2 illustrates how a connected vehicle Maintenance and Fleet-Management System application could work.

Chapter 4 Concepts for the Proposed Applications

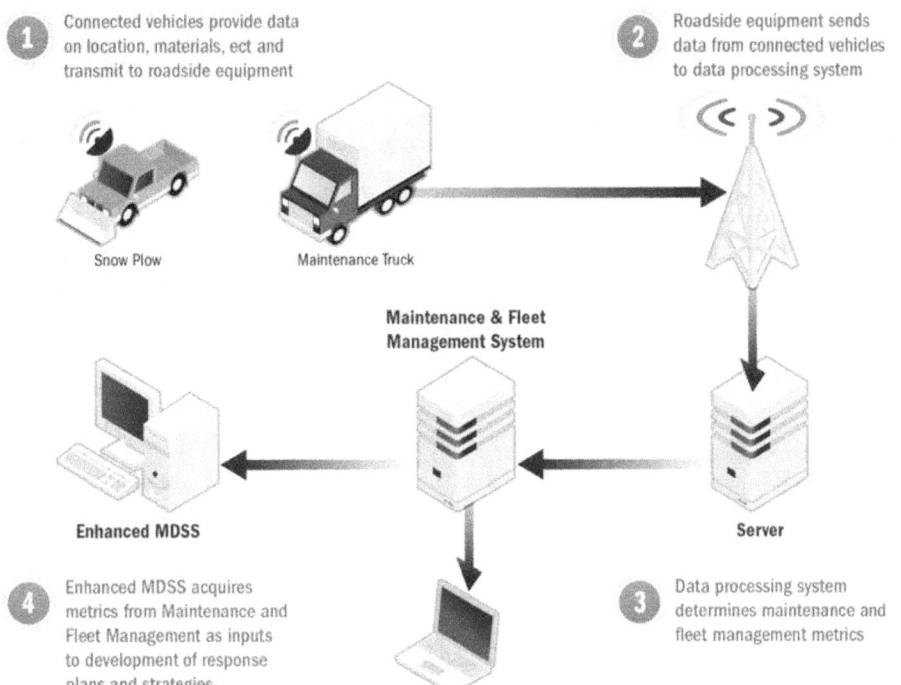

Figure 2: Schematic of Maintenance and Fleet-Management System with Connected Vehicle Road Weather Information

This application will consist of a series of subsystems:

Data Acquisition Subsystem

This subsystem is made up of the connected vehicles, with their associated onboard equipment and the necessary roadside infrastructure. In this concept, connected vehicle data is collected both from vehicles used during winter maintenance and from other maintenance vehicles and equipment used year round. It is assumed that vehicle diagnostic data will be acquired through a CANBus connection. The status of other equipment and information on materials will be gathered via additional specialized sensors installed on the vehicles.

Data Processing Subsystem

Connected vehicle and other sensor data will be communicated via data backhaul to a data processing system. Outputs from this system will be inputs to the Maintenance and Fleet-Management System. When appropriate, the Maintenance and Fleet-Management System will generate data outputs that will be assimilated in the back-end processors of the Enhanced-MDSS for use in developing response plans and treatment strategies.

User Interface System

State and local agency maintenance personnel will interact with the Maintenance and Fleet-Management Systems and, when appropriate during winter maintenance activities, the Enhanced-MDSS when they are using the connected vehicle data described in this concept in a manner similar to the existing systems. Additional decision support tools may be required within the Enhanced-MDSS that use the information provided by the Maintenance and Fleet-Management System.

Operational Policies and Constraints

Data Availability

The effectiveness of this application is predicated on the availability of connected vehicle information. This assumes a willingness of state and local agencies to deploy connected vehicle devices and other specialty sensors into their maintenance vehicles.

MDSS Enhancements

If the existing MDSS is desired as part of this concept, the MDSS will need to undergo enhancements to be able to assimilate and use the connected vehicle information acquired via the Maintenance and Fleet-Management System in its operation. This will require additional research and development that must be defined and performed.

Modes of Operation

The typical modes of operation for the Maintenance and Fleet-Management System concept are—

Normal Mode. In the normal operating mode, the system will be available in advance of and during all maintenance activities, with all designed functionality available.

Degraded Mode. In this mode, some functions are not working properly or may not be available, which could result from many different situations. In the event that connected vehicle data is not available or the Maintenance and Fleet-Management System or the Enhanced-MDSS is not functioning, the remaining system components will operate in a stand-alone manner.

Maintenance Mode. During system maintenance, some subsystems and their functionality may not be available. This mode is similar to Degraded Mode, except that during Maintenance Mode it may be possible to bring the subsystems back into operation, if needed.

User Classes and Other Involved Personnel

Maintenance Vehicle and Equipment Operators. The drivers of these vehicles will be passive participants in the collection and communication of the identified connected vehicle data. However, in the version of the concept that includes interaction between the Maintenance and Fleet-Management System and the Enhanced-MDSS, these vehicle operators will be intended recipients of the information that the Enhanced-MDSS generates. Operators of these vehicles will interact with appropriate in-vehicle devices to receive instructions on their actions during winter weather events.

Maintenance Personnel. This group of users will interact with both the Maintenance and Fleet-Management System and, when appropriate, the Enhanced-MDSS. They will use the outputs of the Maintenance and Fleet-Management System to support a variety of planning, scheduling, and purchasing activities. They will also use the decision support tools available through the Enhanced-

MDSS and direct the actions of the snowplow and other maintenance vehicle operators based on the system outputs.

Support Environment

The Maintenance and Fleet-Management System concept will operate within the overall CVS. As such, the system requires the deployment of connected vehicle onboard equipment, other specialist sensors, and roadside infrastructure; access to the certificate management entities defined for the CVS; and suitable data communications backhaul.

It is assumed that the systems required to acquire and assimilate connected vehicle data into a Maintenance and Fleet-Management System will be deployed coincident with the data processing and communications systems required to operate the existing system within a state or local government facility. However, it is acknowledged that automating the data input process from connected vehicle sources to existing Maintenance and Fleet-Management Systems may present challenges and require specialist development capabilities. Appropriate systems administrators, system maintenance, and IT personnel will be required once the system is operational.

Weather-Responsive Traffic-Management Strategies

Description of the Proposed Weather-Responsive VSL Application

Earlier in this report, the concept of weather-responsive traffic management is discussed. Work underway elsewhere in the connected vehicle program is addressing several traffic-management applications in the areas of V2I safety and dynamic mobility, including signal and stop sign violations, speed harmonization, queue warning, and cooperative adaptive cruise control. This work acknowledges the benefits of integrating road weather information into the applications.

One area that is not being considered elsewhere is VSL for ATDM. This concept describes a Weather-Responsive VSL. VSL systems provide real-time information on appropriate speeds for current conditions and warn drivers of coming road conditions. VSL systems are gaining particular attention for work zone safety management. In this application, the systems consist of multiple roadside monitoring and display trailers, each independently powered and controlled. Each speed trailer uses detectors to measure traffic speed and roadway conditions. A local processor assimilates this information along with other inputs, such as nature and duration of roadwork activity, to determine the appropriate advisory speed or speed limit, which is displayed on a trailer-mounted variable speed sign. The posted speeds can vary throughout the work zone. Opportunities also exist to display variable speed information on infrastructure-mounted DMS or on in-vehicle signing systems.

CVSs provide opportunities to enhance the operation of VSL systems and dramatically improve work zone safety during severe weather events. Additional road weather information can be gathered from connected vehicles and used in algorithms to refine the posted speed advisories or limits to reflect prevailing weather and road conditions. Figure 3 provides a schematic of how a Weather-Responsive VSL could operate in a work zone.

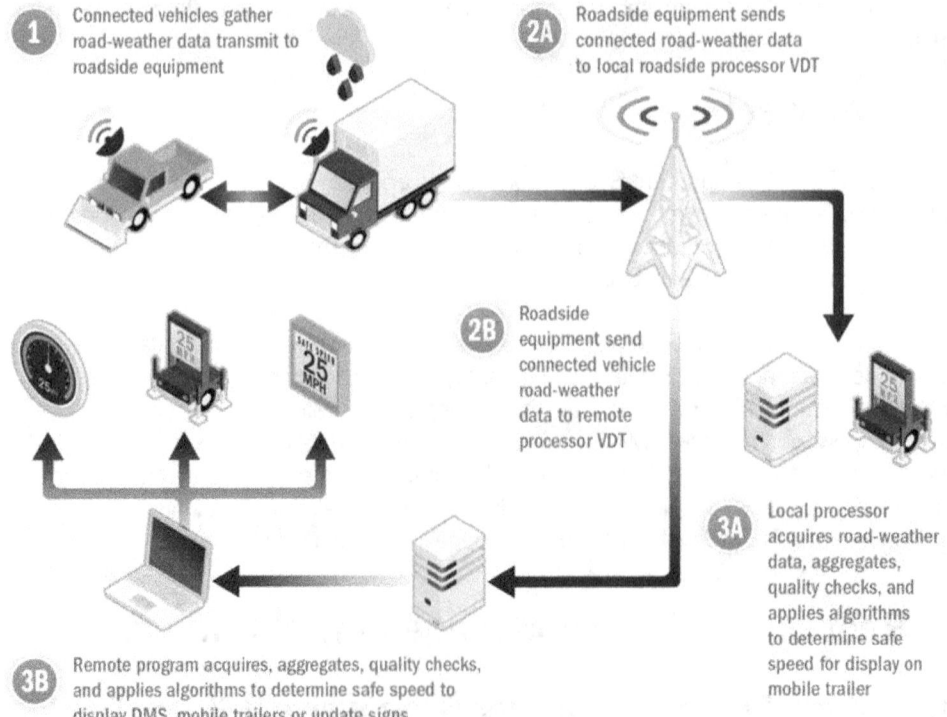

Figure 3: Schematic of Weather-Responsive VSL System

The application will consist of a series of subsystems:

Data Acquisition Subsystem

This subsystem is made up of the connected vehicles with their associated onboard equipment and the appropriate infrastructure at the roadside. Connected vehicles in this application are anticipated to be primarily vehicles operated by the general public and commercial entities (i.e., passenger cars and trucks). It is assumed that passenger cars and commercial trucks will provide data elements specified in BSM Parts 1 and 2 (including the weather-related data elements in BSM Part 2).

Data Processing Subsystem

Two potential alternatives are illustrated for this subsystem. The first assumes that data processing occurs primarily at the roadside using systems installed coincident with portable speed limit sign trailers. This approach may be most suitable for short-lived or mobile work zones (such as paving or striping operations) or in areas where a data communications backhaul capability is not readily available.

The second alternative assumes that data acquired at the roadside are communicated to a remote facility (such as a maintenance shed or a traffic operations center [TOC]) via backhaul. Data processing is performed using systems installed at the remote facility. This deployment approach may be more suitable for long-term construction projects where a semipermanent speed signing infrastructure is installed or for broader VSL applications where speed limit information will be displayed routinely on freeway DMSs or through in-vehicle signing systems.

In both cases, the acquired connected vehicle road weather data is processed using a VDT to generate weather and road condition variables, such as—

- Weather variables, such as air temperature, barometric pressure, and dew point

- Road weather variables, such as pavement temperature and friction

- Other variables, such as average vehicle speeds, ABS activation events, and vehicle stability and traction control activation events

- Inferred variables (from VDT algorithms), such as slickness, visibility, and precipitation rate and type.

Outputs from the VDT along with other traffic data, other atmospheric weather information, surface weather observations, and work zone characteristics are used in a new speed limit selection algorithm to develop recommendations on appropriate travel speeds under the prevailing traffic, weather, and road conditions.

Information Display Subsystems

In the concept where data processing occurs at the roadside, the information display subsystem will comprise the local, mobile speed limit display trailers commonly used in work zones. Where data processing occurs in a remote facility, VSL information can be displayed on any appropriate signage for which communications capabilities exist. This may include freeway and arterial DMS. After a suitable connected vehicle field infrastructure is deployed, the information display system could also include in-vehicle signing devices.

Description of the Proposed Weather-Responsive Signalized Intersection Application

A further area of WRTM that provides a particular opportunity for the use of road weather connected vehicle data is in signalized intersection control during deteriorating and adverse weather conditions. Research shows that benefits may accrue to traffic at signalized intersections during inclement weather through the use of the following strategies:

- Changes to the clearance interval (the yellow and all-red period) at signalized intersections

- Changes to green intervals to accommodate start-up lost time and longer discharge headways particularly on snow, slush, or ice covered pavements

- Development and implementation of special signal timing or coordination plans for use during adverse weather events, particularly under snow and ice conditions.

Road weather information collected by CVSs can be processed and used in algorithms associated with a local traffic signal controller to adjust intervals in response to current road and weather conditions. Information from connected vehicles can also be transmitted to remote agency locations where it is used in algorithms that provide recommendations to traffic managers to implement special signal timing plans more appropriate for prevailing road weather conditions. Figure 4 provides a schematic of how a Weather-Responsive Signalized Intersection could operate.

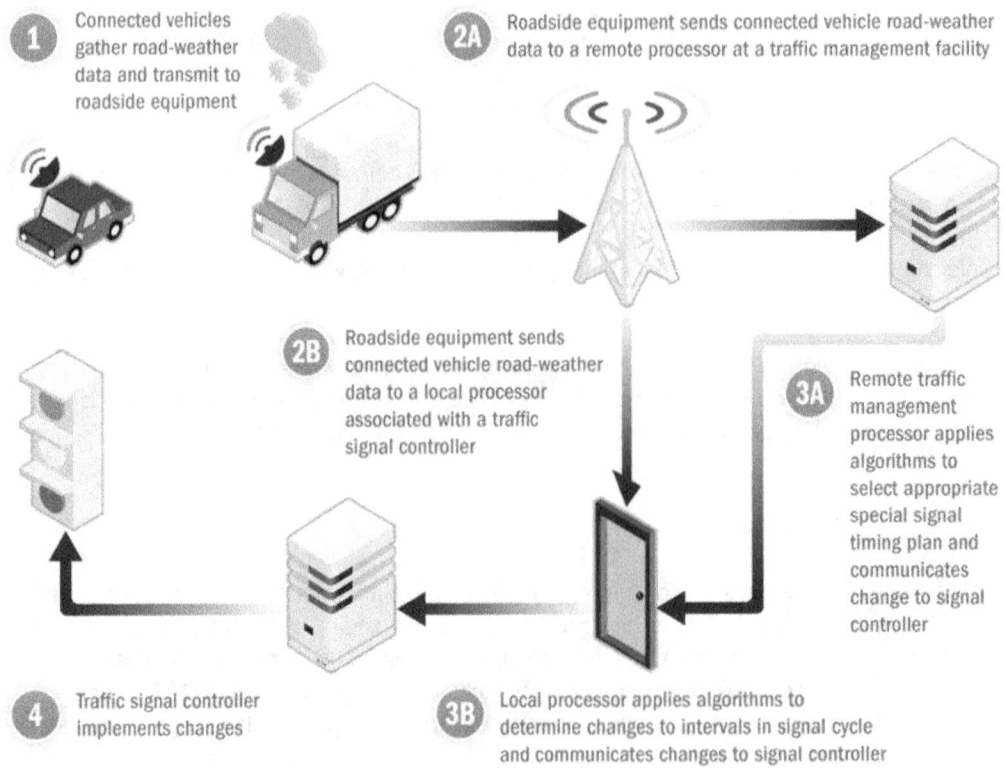

Figure 4: Schematic of Weather-Responsive Signalized Intersection System

The application will consist of a series of subsystems:

Data Acquisition Subsystem

This subsystem is made up of the connected vehicles with their associated onboard equipment and the appropriate infrastructure at the roadside. Connected vehicles in this application are anticipated to be primarily vehicles operated by the general public and commercial entities (i.e., passenger cars and trucks). It is assumed that passenger cars and commercial trucks will provide data elements specified in BSM Parts 1 and 2 (including the weather-related data elements in BSM Part 2).

Data Processing Subsystem

Two potential alternatives are illustrated for this subsystem. The first assumes that data processing occurs at the roadside using systems installed coincident with a traffic signal controller at a signalized intersection. This approach is considered suitable for small changes to green intervals or clearance intervals during a signal cycle to accommodate changes in vehicle speeds and headways caused by poor weather-related pavement conditions.

The second alternative assumes that data acquired at the roadside is communicated to a remote facility (most likely a traffic-management facility) via backhaul. Data processing is performed using systems installed at the remote facility. This deployment approach will be suitable for identifying the appropriate special signal timing plan and implementing that plan for the duration of a weather event.

In both cases, the acquired connected vehicle road weather data is processed using a VDT to generate weather and road condition variables, such as—

- Weather variables, such as air temperature, barometric pressure, and dew point
- Road weather variables, such as pavement temperature and friction
- Other variables, such as average vehicle speeds, ABS activation events, and vehicle stability and traction control activation events
- Inferred variables (from VDT algorithms), such as slickness, visibility, and precipitation rate and type.

Outputs from the VDT along with other traffic data, other atmospheric weather information, surface weather observations, and information about the signalized intersection configuration and signal timing plans are used in new algorithms to develop appropriate signal timing changes under the prevailing traffic, weather, and road conditions.

Information Display Subsystems

In the concept where data processing occurs at a remote facility, a visual display will provide information to traffic managers for use in their selection of the signal timing plan. In both scenarios, the form of information display to drivers is the existing traffic signal head.

Operational Policies and Constraints

Data Availability

The effectiveness of this application is predicated on the availability of connected vehicle road weather information. This assumes a broad penetration of connected vehicle onboard equipment into the national vehicle fleet and, in one of the suggested deployment approaches, the availability of an appropriate roadside and data backhaul infrastructure. Additional research will clarify the levels of vehicle fleet penetration required to generate sufficient data for the application to work effectively.

In addition, in this application, data is especially desired in advance of predicted winter storms and during other severe winter weather events. During these periods, drivers are encouraged and inclined to avoid travel, which may affect the availability of the required road weather data.

VDT Implementation

This application will require participating agencies or their contractors to implement and operate the VDT. The VDT is currently in a development phase; therefore, the impacts of this requirement are currently unknown.

Enhancement of Mobile Speed Limit Trailers

This application, in one of the recommended deployment approaches, will require state and local agencies to equip mobile speed limit trailers used in work zones with the appropriate roadside data-acquisition components and the systems needed for data processing by the VDT and the speed limit selection algorithm.

Enforceability of Variable Speed Limits

Depending on the proposed deployment approach, enforcement of VSLs may require state or local legislation to be enacted.

Enhancement of Traffic Signal Controllers

This application will require state and local agencies to equip traffic signal controllers with the appropriate roadside data-acquisition components and the systems needed for data processing by the VDT and the algorithm that will identify required changes to the signal cycle.

Modes of Operation

The typical modes of operation for the WRTM concepts are—

Normal Mode. In the normal operating mode, the WRTM applications will be available in advance of and during all appropriate severe weather events with all designed functionality available. Details of Normal Mode of operation may vary depending on the operational policies of the agency. For example, in the VSL application, during periods when the work zone is inactive, some agencies may wish to display no speed limit information irrespective of the prevailing traffic, weather, and road conditions.

Degraded Mode. In this mode, some functions are not working properly or may not be available. This could result from many different situations. In the event that connected vehicle road weather data are not available or the VDT is not functioning, the system will operate in the manner of existing work zone speed limit displays.

Maintenance Mode. During system maintenance, some subsystems and their functionality may not be available. This mode is similar to Degraded Mode, except that during Maintenance Mode it may be possible to bring the subsystems back into operation, if needed.

User Classes and Other Involved Personnel

Vehicle Operators. From a data-delivery standpoint, vehicle operators will be passive participants in the operation of the WRTM applications. While operating their vehicles, onboard equipment will collect connected vehicle road weather information and communicate this information to appropriate roadside equipment. From an information-dissemination standpoint, all vehicle operators (irrespective of whether they were a data provider) will be recipients of the information generated either by the Weather-Responsive VSL and displayed on the work zone speed limit display trailers or by the Weather-Responsive Signalized Intersection application and displayed on the signal head in the conventional manner.

Work Zone Personnel. In the Weather-Responsive VSL application, this group of users will be responsible for the correct placement of speed limit sign trailers and for activation of the roadside data processing system in situations where speed limit information is displayed only when the work zone is active.

Traffic Operations Personnel. This group of users will interact with the Weather-Responsive VSL in the deployment scenario, where data processing takes place at a remote TOC. They will be presented recommendations on appropriate speeds by roadway segment under the prevailing conditions and will make decisions to post speed limit information on the DMS under their control. This group of users will also interact with the Weather-Responsive Signalized Intersection application in the deployment scenario where data processing takes place at a remote traffic management facility. They will be presented recommendations on appropriate special signal timing plans under the prevailing conditions and will make decisions to implement the plans on the signal systems under their control.

Support Environment

The WRTM concepts will operate within the overall CVS. As such, the WRTM applications require the deployment of connected vehicle onboard equipment and roadside infrastructure, access to the certificate management entities defined for the CVS, and suitable data communications backhaul where data processing occurs at a remote facility.

It is assumed that the data processing and communications systems operating the Weather-Responsive VSL will be deployed on mobile work zone speed limit sign trailers or within a state or local government facility, and the data processing and communications systems operating the Weather-Responsive Signalized Intersections will be deployed in association with existing or upgraded traffic signal controllers or within a state or local government facility. Appropriate systems administrators, system maintenance (including field maintenance personnel for the trailer-based equipment), and IT personnel will be required.

A suitable communications infrastructure, in common with existing traffic operations practices, will be required for the personnel using the WRTM applications.

Motorist Advisories and Warnings

Description of the Proposed Application

Motorists currently have access to a variety of traveler and weather information from multiple sources and providers and through many different media. State DOTs typically provide information on significant traffic incidents and delays; work zones; and the impacts of severe weather events, such as road closures caused by winter storms or flooding. Travelers can access this information through 511 systems or other phone-based hotlines or agency websites or see it displayed on roadway infrastructure such as DMS, or—increasingly—through social media. Public agencies usually provide this information to the traditional media outlets, as well, so travelers can obtain information from radio and television broadcasts. Atmospheric weather information is similarly distributed by the public sector: Travelers can access NWS Watches, Warnings, Statements, and Advisories through a variety means, and broadcast media outlets use NWS Doppler Radar feeds to create weather forecasts.

The private sector also offers several sources, tools, and services to provide motorist with weather and travel information. Increasingly, the private sector is packaging traveler information with consumer navigation products or providing the information through applications on smart phones or other personal mobile devices. Business models vary from free applications, from those with a small, one-time fee to those that are subscription based.

In all cases, the value of the information provided to the traveler is directly related to the breadth and quality of the data collection capabilities. Within this environment, information on segment-specific weather and road conditions is not well represented, even though surveys suggest that this information is considered to be of significant importance to travelers.

The ability to gather road weather information from connected vehicles will dramatically change this situation. Two information loops can be envisioned in this application: The first emphasizes gathering and disseminating spot warnings and advisories directly to individual motorists in the fastest possible means, while the second focuses on the integration of road weather information into a broader set of advisories but over a longer time period. Information on deteriorating road and weather conditions on specific roadway segments can be pushed to travelers through a variety of means as alerts and advisories within a few minutes. In combination with observations and forecasts from other sources and with additional processing, medium-term advisories of the next 2–12 hours to long-term advisories for more than 12 hours into the future can also be provided to motorists. In both situations, the connected vehicle information provides the opportunity to create advisories and warnings with greater temporal and geographic resolution than is otherwise currently available.

Figure 5 provides a schematic of how a Road-Weather Motorist Advisory and Warning System could operate.

Chapter 4 Concepts for the Proposed Applications

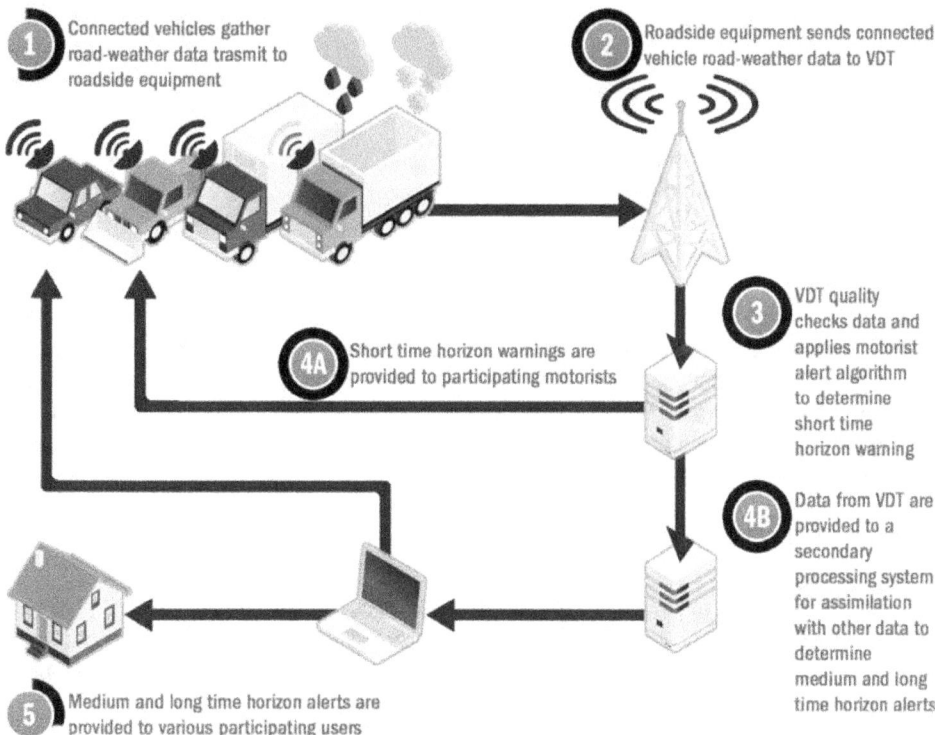

Figure 5: Schematic of Road Weather Motorist Advisory and Warning System

The application will consist of a series of subsystems:

Data Acquisition Subsystem

This subsystem is made up of the connected vehicles with their associated onboard equipment and the necessary roadside infrastructure. Two classes of connected vehicle are anticipated for the application: vehicles operated by the general public and commercial entities (including passenger cars and trucks) and specialty vehicles and public fleet vehicles (such as snowplows, maintenance trucks, and other agency pool vehicles). It is assumed that passenger cars and commercial trucks will provide data elements specified in BSM Parts 1 and 2 (including the weather-related data elements in BSM Part 2), while agency-controlled vehicles will provide these data elements and, optionally, additional data elements from specialty sensors installed on selected vehicles (e.g., sensors to measure salinity at the roadway surface).

Data Processing Subsystem

Connected vehicle road weather data will be communicated via data backhaul to a remote VDT or to roadside VDT, where connected vehicle road weather data is communicated from vehicles via DSRC. The VDT will process the raw data and generate road segment-based data outputs. These data outputs will be provided to a road weather motorist alerts algorithm to generate short time horizon alerts that will be pushed to user systems and available to commercial service providers.

Outputs from the VDT will include—

- Weather variables, such as air temperature, barometric pressure, and dew point

- Road weather variables, such as pavement temperature, friction, and salinity

- Other variables, such as average vehicle speeds, ABS activation events, and vehicle stability and traction control activation events

- Inferred variables (from VDT algorithms), such as slickness, visibility, and precipitation rate and type.

VDT output data will supplement other data sources (such as data from national weather models and from surface weather and road weather observation systems) and will be assimilated in back-end processors for use in the various weather and pavement temperature models. These outputs will be available to a variety of public and private-sector users for use in tools that generate medium- and long-horizon alerts and warnings.

Information Generation Subsystem

Data outputs from the VDT will be available to other information processing systems that may reside in public agencies or be operated by commercial service providers. These systems are intended to produce tailored information content for the various user systems, particularly for medium- and long-horizon motorist alerts and warnings. Additional decision support tools may also be developed for state and local agency traffic and maintenance operations personnel who use the detailed roadway segment-specific data that is newly available from the connected vehicle data sources and the VDT.

User Interface Subsystems

Outputs from the road weather motorist alerts algorithm (i.e., the short time horizon alerts) and outputs from other information generation subsystems will be provided in a manner that makes the information accessible through as many user interfaces (UI) as possible. These will likely include phones and other personal mobile devices, websites, infrastructure-based displays, and traditional broadcast media.

Operational Policies and Constraints

Data Availability

The effectiveness of this application is predicated on the availability of connected vehicle road weather information. This assumes a broad penetration of connected vehicle onboard equipment into the national vehicle fleet and the availability of an appropriate roadside and data backhaul infrastructure. It further assumes the willingness of state and local agencies to deploy connected vehicle devices and potentially other specialty sensors into the vehicles under their control. Additional research may clarify the levels of agency and general fleet penetration required to generate sufficient data for the application to work effectively.

In addition, in this application, data is especially desired in advance of predicted winter storms and during other severe winter weather events. During these periods, drivers are encouraged and inclined to avoid travel, which may affect the availability of the required road weather data.

VDT Implementation

This application will require participating agencies to implement and operate the VDT. The VDT is currently in a development phase; therefore, the impacts of this requirement are currently unknown.

Algorithm and Information Processing System Development

As described, this concept relies on the development of new algorithms to rapidly analyze connected vehicle road weather data to generate short time horizon alerts plus the systems and communications required to push these alerts to users. In addition, the concept describes other information processing systems that will generate and distribute medium- and long-horizon motorist advisories and alerts. This will require additional research and development that must be defined and performed.

The concept further assumes that information processing and the development of new information products will occur in both the public and private sectors. It is likely that the development of tailored information content by the private sector will be driven by market forces rather than a desire to disseminate weather and road condition information to improve roadway safety for the public good.

Interfaces to Other Systems

The concept assumes that the short-, medium-, and long-horizon motorist alerts and advisories will be delivered through a variety of systems, including public-sector websites, 511 and other phone-based information systems, DMSs, and social media. Suitable interfaces will need to be developed to existing systems of these types that reside in the public sector.

Deployment Coverage

A sufficiently dense network of roadside equipment with adequate geographic coverage will be required to gather connected vehicle road weather data that are effective for the concept. This will be particularly important in areas of complex terrain or where information on short roadway segments is desired.

Modes of Operation

The typical modes of operation for the Road Weather Motorist Advisory and Warning System concept are—

Normal Mode. In the normal operating mode, the system will be available during all adverse weather events, with all designed functionality available. The system will provide short time horizon alerts to system users, with minimal delay from the time of data acquisition.

Degraded Mode. In this mode, some functions are not working properly or may not be available, which could result from many different situations. In the event that connected vehicle road weather data are not available or the VDT is not functioning, the system will be unable to provide short time horizon motorist alerts. Medium and long time horizon alerts may be delivered using data available from only national weather models and fixed and remote sensor systems.

Maintenance Mode. During system maintenance, some subsystems and their functionality may not be available. During Maintenance Mode, it may be possible to bring the subsystems back into operation, if needed.

User Classes and Other Involved Personnel

Vehicle Operators. From a data-delivery standpoint, vehicle operators will be passive participants in the operation of the Road Weather Motorist Advisory and Warning System. While operating their vehicles, onboard equipment with collect connected vehicle road weather information and will communicate this information to appropriate roadside equipment. From an information-dissemination standpoint, all vehicle operators (irrespective of whether they were a data provider) may be recipients of the information the system generates depending on the user devices available to them.

Traffic and Maintenance Operations Personnel. These groups of users may interact with the Road Weather Motorist Advisory and Warning System. They may be required to disseminate information via the traffic-management infrastructure or to maintenance personnel in the field.

Public Agency Traveler Information Providers. This group of users may be required to disseminate information from the Road Weather Motorist Advisory and Warning System to the public via 511 or other phone-based systems or through agency websites.

Commercial Service Providers. This group of users will receive information from the data processing subsystems and will develop information products tailored to the needs of their customers.

Support Environment

The Road Weather Motorist Advisory and Warning System concept will operate within the overall CVS. As such, the Enhanced-MDSS requires the deployment of connected vehicle onboard equipment and roadside infrastructure, access to the certificate management entities defined for the CVS, and suitable data communications backhaul.

It is assumed that the systems operating the VDT and other information processing systems will be deployed within a state or local government facility. Appropriate systems administrators, system maintenance, and IT personnel will be required.

Suitable communications infrastructure, in common with the existing traveler information and traffic operations systems, will be required for the dissemination of alerts and advisories to users.

Information for Freight Carriers

Description of the Proposed Application

This application can be considered a special case of the Road Weather Motorist Advisory and Warning System described in the previous section. Truck drivers have similar access to the variety of traveler information systems that are available to all road users. However, the available traveler information options are invariably intended for use by passenger car drivers. The limitations of the

Chapter 4 Concepts for the Proposed Applications

existing systems with respect to the type and quality of information provided have particular impacts on motor carriers.

Prevailing and forecast weather conditions and the impacts of weather on the roadways are especially significant to freight carriers. Drivers must be conscious of current roadway conditions to safely operate the vehicles and must be aware of approaching weather events or deteriorating conditions to plan their hours of service and to seek suitable truck parking locations. Because of the nature of many truck trips, multistate information is also especially important.

In the event that a particular roadway segment becomes impassable because of weather conditions, truck drivers face greater challenges in rerouting. Truck drivers must coordinate with their dispatchers to ensure that an alternative route is suitable, considering highway weight restrictions, bridge height restrictions, or geometric issues such as tight turning radii. Dispatchers must also consider other operational factors, such as delivery schedules, when making decisions to delay a trip or reroute because of weather events.

The ability to gather road weather information from connected vehicles will significantly improve the ability of freight shippers to plan and respond to the impacts of severe weather events and poor road conditions. Information on deteriorating road and weather conditions on specific roadway segments can be pushed to both truck drivers and their dispatchers through a variety of means as alerts and advisories with low latency. In combination with observations and forecasts from other sources and with additional processing, medium- to long-term advisories can also be provided to dispatchers to support routing and scheduling decisions. Because these decisions must consider a variety of other factors, such as highway and bridge restrictions, hours-of-service limitations, parking availability, delivery schedules, and—in some instances—the permits the vehicle holds, it is envisioned that the motor carrier firms or their commercial service providers will develop and operate the systems that use the road weather information generated through this concept. This connected vehicle information can also support decisions made by state agencies relating to the temporary suspension of commercial vehicle permits because of prevailing weather and road conditions. Up-to-date information on actual conditions at high resolution may allow temporary restrictions to be lifted in a more timely manner.

Figure 6 provides a schematic of how a Road Weather Advisory and Warning System for Freight Carriers could operate.

Chapter 4 Concepts for the Proposed Applications

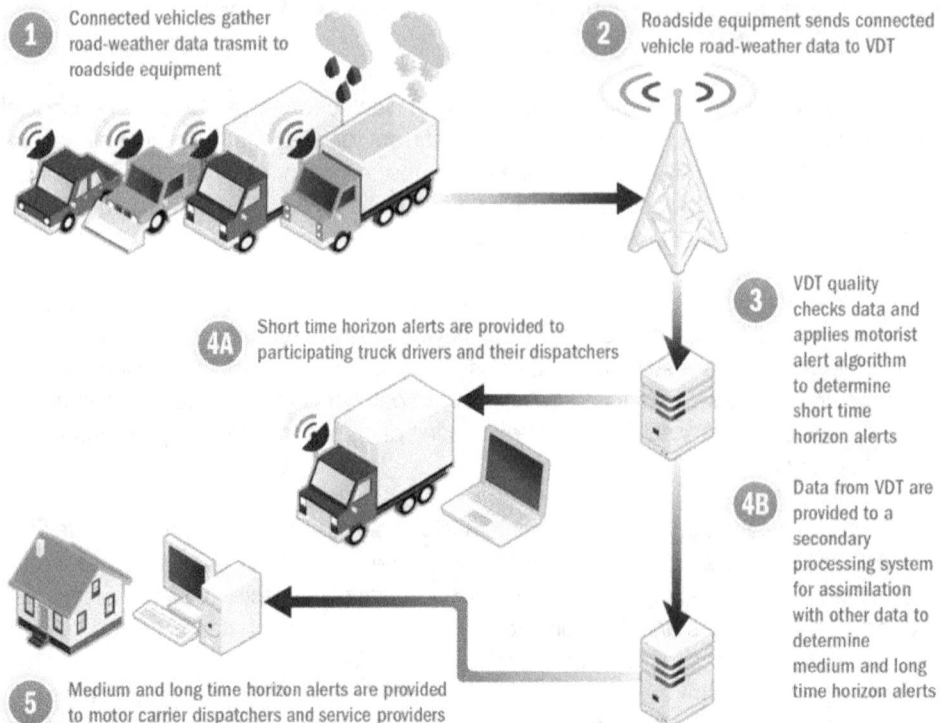

Figure 6: Schematic of Road Weather Advisory and Warning System for Freight Carriers

The application will consist of a series of subsystems:

Data Acquisition Subsystem

This subsystem is made up of the connected vehicles, with their associated onboard equipment and the necessary roadside infrastructure. Two classes of connected vehicle are anticipated for the application: vehicles operated by the general public and commercial entities (including passenger cars and trucks) and specialty vehicles and public fleet vehicles (such as snowplows, maintenance trucks, and other agency pool vehicles). It is assumed that passenger cars and commercial trucks will provide data elements specified in BSM Parts 1 and 2 (including the weather-related data elements in BSM Part 2), while agency-controlled vehicles will provide these data elements and, optionally, additional data elements from specialty sensors installed on selected vehicles (e.g., sensors to measure salinity at the roadway surface).

Data Processing Subsystem

Connected vehicle road weather data will be communicated via data backhaul to a VDT. The VDT will process the raw data and generate road segment-based data outputs. These data outputs will be provided to a road weather motorist alert algorithm to generate short time horizon alerts that will be pushed to truck drivers and their dispatchers.

Outputs from the VDT will include—

- Weather variables, such as air temperature, barometric pressure, and dew point

- Road weather variables, such as pavement temperature, friction, and salinity

- Other variables, such as average vehicle speeds, ABS activation events, and vehicle stability and traction control activation events

- Inferred variables (from VDT algorithms), such as slickness, visibility, and precipitation rate and type.

In addition to the VDT outputs, the short time horizon alerts that are pushed to truck drivers will include high-wind advisories. This information will be acquired from other fixed and remote observation systems and will be provided with as much geographic precision as possible.

VDT output data will supplement other data sources (such as data from national weather models and from surface weather and road weather observation systems) and will be assimilated in back-end processors for use in the various weather and pavement temperature models. The outputs from these processes will be available to motor carrier dispatchers and their service providers for use in tools that generate medium- and long-horizon alerts and warnings and that may determine route and schedule adjustments for individual vehicles based on the weather and road conditions.

Information Generation Subsystem

Data outputs from the VDT will be available to other information processing systems that may reside in freight shipper facilities or be operated by commercial service providers. These systems will be intended to produce tailored information content for the various user systems, particularly for medium- and long-horizon alerts and warnings.

User Interface Subsystems

Outputs from the road weather motorist alerts algorithm (i.e., the short time horizon alerts) and outputs from other information generation subsystems will be provided in a manner that makes the information accessible through UIs that are appropriate for the truck cab environment or trucking firm dispatcher.

Operational Policies and Constraints

Data Availability

The effectiveness of this application is predicated on the availability of connected vehicle road weather information. This assumes a broad penetration of connected vehicle onboard equipment into the national vehicle fleet and the availability of an appropriate roadside and data backhaul infrastructure. It further assumes the willingness of state and local agencies to deploy connected vehicle devices and potentially other specialty sensors into the vehicles under their control. Additional research may clarify the levels of agency and general fleet penetration required to generate sufficient data for the application to work effectively.

In addition, in this application, data are especially desired in advance of predicted winter storms and during other severe winter weather events. During these periods, drivers are encouraged and inclined to avoid travel, which may affect the availability of the required road weather data.

VDT Implementation

This application will require participating agencies to implement and operate the VDT. The VDT is currently in a development phase; therefore, the impacts of this requirement are currently unknown.

Algorithm and Information Processing System Development

As described, this concept relies on the development of new algorithms to rapidly analyze connected vehicle road weather data to generate short time horizon alerts plus the systems and communications required to push these alerts to truck drivers and dispatchers. In addition, the concept describes other information processing systems that will generate and distribute medium- and long-horizon motorist advisories and alerts. This will require additional research and development that must be defined and performed.

The concept further assumes that information processing and the development of new information products will occur primarily in the private sector. It is likely that the development of these systems by the private sector will be driven by market forces. An analysis of the needs of the motor carrier industry for these products and the l kelihood of their development by commercial service providers may be appropriate prior to additional development of this concept by USDOT.

Interfaces to Other Systems

The concept assumes that the short-, medium-, and long-horizon alerts and advisories will be delivered through a variety of systems native to trucks and trucking dispatch offices. Suitable interfaces will need to be developed to these existing systems.

Deployment Coverage

A sufficiently dense network of roadside equipment with adequate geographic coverage will be required to gather connected vehicle road weather data that is effective for the concept. This will be particularly important in areas of complex terrain or where information on short roadway segments is desired.

Modes of Operation

The typical modes of operation for the Road Weather Advisory and Warning System for Freight Shippers concept are—

Normal Mode. In the normal operating mode, the system will be available during all adverse weather events, with all designed functionality available. The system will provide short time horizon alerts to system users with minimal delay from the time of data acquisition.

Degraded Mode. In this mode, some functions are not working properly or may not be available, which could result from many different situations. In the event that connected vehicle road weather data are not available or the VDT is not functioning, the system will be unable to provide short time horizon motorist alerts. Medium and long time horizon alerts may be delivered using data available from only national weather models and fixed and remote sensor systems.

Maintenance Mode. During system maintenance, some subsystems and their functionality may not be available. During Maintenance Mode, it may be possible to bring the subsystems back into operation, if needed.

User Classes and Other Involved Personnel

Vehicle Operators. From a data-delivery standpoint, vehicle operators will be passive participants in the operation of the Road Weather Advisory and Warning System for Freight Shippers. While operating their vehicles, onboard equipment will collect connected vehicle road weather information and communicate this information to appropriate roadside equipment. From an information-dissemination standpoint, vehicle operators (other than truck drivers, who are described as a separate class below) will not be recipients of the information that this system generates.

Truck Drivers. From a data-delivery standpoint, truck drivers will be passive participants in the operation of the Road Weather Advisory and Warning System for Freight Shippers. While operating their vehicles, onboard equipment will collect connected vehicle road weather information and communicate this information to appropriate roadside equipment. From an information-dissemination standpoint, participating truck drivers will be recipients of the information that this system generates.

Truck Dispatchers. Dispatchers will be recipients of all of the information that this system generates. Dispatchers will use the information to advise truck drivers or to reroute and reschedule truck trips.

Commercial Service Providers. This group of users will receive information from the data processing subsystems and will develop information products tailored to the needs of their customers.

Support Environment

The Road Weather Advisory and Warning System for Freight Shippers concept will operate within the overall CVS. As such, the system requires the deployment of connected vehicle onboard equipment and roadside infrastructure, access to the certificate management entities defined for the CVS, and suitable data communications backhaul.

It is assumed that the systems operating the VDT and other information processing systems will be deployed within a state or local government facility. Appropriate systems administrators, system maintenance, and IT personnel will be required. Suitable communications infrastructure will be required for the dissemination of alerts and advisories to truck drivers, dispatchers, and commercial service providers.

Information and Routing Support for Emergency Responders

Description of the Proposed Application

Emergency responders, including ambulance operators, paramedics, and fire and rescue organizations, have a compelling need for the short, medium, and long time horizon road weather alerts and warnings that have been described for the two previous concepts. This information can help drivers safely operate their vehicles during severe weather events and under deteriorating road conditions.

In addition, however, emergency responders have a particular need for information that affects their dispatching and routing decisions. Information on weather-affected travel routes—especially road or lane closures caused by snow, flooding, and wind-blown debris—is particularly important. Low-latency road weather information from connected vehicles for specific roadway segments together with information from other surface weather observation systems, such as flooding and high winds, will be used to determine response routes, calculate response times, and influence decisions to hand off an emergency call from one responder to another responder in a different location.

Figure 7 provides a schematic of how a Road Weather-Sensitive Emergency Responders System could operate.

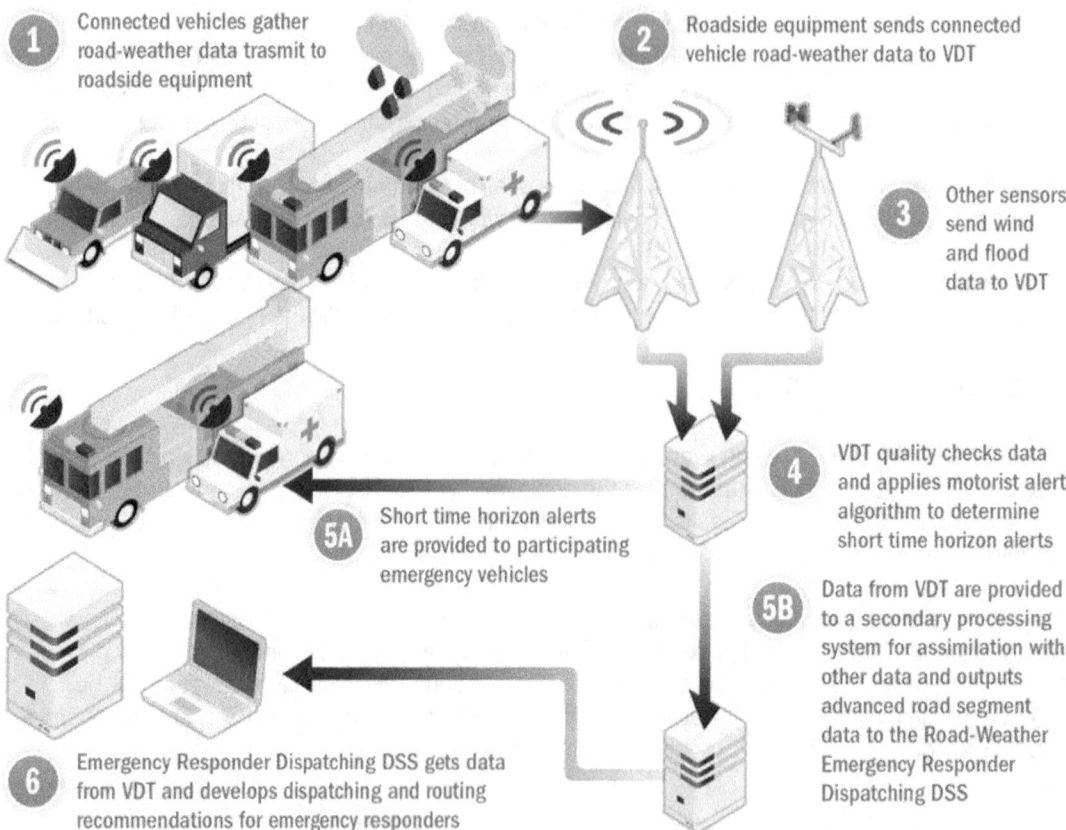

Figure 7: Schematic of Road Weather Emergency Responder Dispatching DSS

The application will consist of a series of subsystems:

Data Acquisition Subsystem

This subsystem is made up of the connected vehicles, with their associated onboard equipment and the necessary roadside infrastructure. Three classes of connected vehicle are anticipated for the application: vehicles operated by the general public and commercial entities (including passenger cars and trucks), emergency vehicles (including ambulances and fire trucks), and specialty vehicles and public fleet vehicles (such as snowplows, maintenance trucks, and other agency pool vehicles). It is assumed that passenger cars and commercial trucks will provide data elements specified in BSM Parts 1 and 2 (including the weather-related data elements in BSM Part 2), while emergency and public agency-controlled vehicles will provide these data elements and, optionally, additional data elements from specialty sensors installed on selected vehicles.

Data Processing Subsystem

Connected vehicle road weather data will be communicated via data backhaul to a VDT. The VDT will process the raw data and generate road segment-based data outputs. These data outputs will be provided to a road weather motorist alerts algorithm to generate short time horizon alerts that will be pushed to emergency vehicle drivers and first-responder dispatchers.

Outputs from the VDT will include—

- Weather variables, such as air temperature, barometric pressure, and dew point

- Road weather variables, such as pavement temperature, friction, and salinity

- Other variables, such as average vehicle speeds, vehicle stops, delays, traffic queue build-ups, ABS activation events, and vehicle stability and traction control activation events

- Inferred variables (from VDT algorithms), such as slickness, visibility, and precipitation rate and type.

In addition to the VDT outputs, the short time horizon alerts that are pushed to emergency vehicle drivers and dispatchers will include information on high winds, standing water, and flooding of roadways. This information will be acquired from other fixed and remote observation systems and will be provided with as much geographic precision as possible.

VDT output data will supplement other data sources (such as data from national weather models and from surface weather and road weather observation systems) and will be assimilated in back-end processors for use in the various weather and pavement temperature models. The outputs from these processes will be passed to a new Road Weather Emergency Responder Dispatching DSS.

Road Weather Emergency Responder Dispatching DSS

Data outputs from the VDT will be available to a new Road Weather Emergency Responder Dispatching DSS that may reside in an emergency responder facility or a transportation agency. This system will be intended to analyze the complex interactions between current and forecast road weather conditions, other current surface weather observations, current roadway traffic conditions (including average traffic speeds and congestion conditions by roadway segment), and the communications needs with road agencies (such as information on which roads are plowed; the status of removing wind-blown debris from roadways; and the presence of incidents, work zones, or other situations causing lane or roadway closures).

User Interface Subsystems

Outputs from the Road Weather Emergency Responder Dispatching DSS will be provided in a manner that makes the information accessble through UIs that are appropriate for the emergency vehicle or emergency responder dispatcher.

Operational Policies and Constraints

Data Availability

The effectiveness of this application is predicated on the availability of connected vehicle road weather information. This assumes a broad penetration of connected vehicle onboard equipment into the national vehicle fleet and the availability of an appropriate roadside and data backhaul infrastructure. It further assumes the willingness of emergency responders and state and local agencies to deploy connected vehicle devices and potentially other specialty sensors into the vehicles under their control. Additional research may clarify the levels of agency and general fleet penetration required to generate sufficient data for the application to work effectively.

In addition, in this application, data are especially desired in advance of predicted winter storms and during other severe winter weather events. During these periods, drivers are encouraged and inclined to avoid travel, which may affect the availability of the required road weather data.

VDT Implementation

This application will require participating agencies to implement and operate the VDT. The VDT is currently in a development phase; therefore, the impacts of this requirement are currently unknown.

Road Weather Emergency Responder Dispatching DSS Development

As described, this concept relies on the development of new algorithms to rapidly analyze connected vehicle road weather data to generate short time horizon alerts plus a new Road Weather Emergency Responder Dispatching DSS. This will require additional research and development that must be defined and performed. An analysis of the needs of the emergency responder community for this system may be appropriate prior to additional development of this concept by USDOT.

Interfaces to Other Systems

The concept assumes that the short time horizon alerts and advisories will be delivered through a variety of systems native to emergency vehicles and emergency responder dispatch facilities. Suitable interfaces will need to be developed to these existing systems.

Deployment Coverage

A sufficiently dense network of roadside equipment with adequate geographic coverage will be required to gather connected vehicle road weather data that is effective for the concept. This will be particularly important in areas of complex terrain or where information on short roadway segments is desired.

Modes of Operation

The typical modes of operation for the Road Weather Emergency Responder Dispatching DSS concept are—

Normal Mode. In the normal operating mode, the system will be available during all adverse weather events, with all designed functionality available. The system will provide short time horizon alerts to system users with minimal delay from the time of data acquisition.

Degraded Mode. In this mode, some functions are not working properly or may not be available, which could result from many different situations. In the event that connected vehicle road weather data is not available or the VDT is not functioning, the system will be unable to provide short time horizon motorist alerts. Medium and long time horizon alerts may be delivered to the Road Weather Emergency Responder Dispatching DSS using data available from only national weather models and fixed and remote sensor systems.

Maintenance Mode. During system maintenance, some subsystems and their functionality may not be available. During Maintenance Mode, it may be possible to bring the subsystems back into operation, if needed.

User Classes and Other Involved Personnel

Vehicle Operators. From a data-delivery standpoint, vehicle operators will be passive participants in the operation of the Road Weather Emergency Responder Dispatching DSS. While operating their vehicles, onboard equipment will collect connected vehicle road weather information and communicate this information to appropriate roadside equipment. From an information-dissemination standpoint, vehicle operators (other than emergency vehicle drivers, who are described as a separate class below) will not be recipients of the information that this system generates.

Emergency Vehicle Drivers. From a data-delivery standpoint, truck drivers will be passive participants in the operation of the Road Weather Emergency Responder Dispatching DSS. While operating their vehicles, onboard equipment will collect connected vehicle road weather information and communicate this information to appropriate roadside equipment. From an information-dissemination standpoint, participating emergency vehicle drivers will be recipients of the information that this system generates.

Emergency Responder Dispatchers. Dispatchers will be recipients of all of the information that this system generates. This group of users will also interact with the new Road Weather Emergency Responder Dispatching DSS. They will use the decision support tools available through the new Road Weather Emergency Responder Dispatching DSS and will direct the handling of emergency calls and the routing of emergency vehicles based on the system outputs. Alternatively, this group of users may interact with a commercial service provider that provides the information from this concept to the emergency responder dispatcher.

Support Environment

The Road Weather Emergency Responder Dispatching DSS concept will operate within the overall CVS. As such, the system requires the deployment of connected vehicle onboard equipment and roadside infrastructure, access to the certificate management entities defined for the CVS, and suitable data communications backhaul.

It is assumed that the systems operating the VDT and the new Road Weather Emergency Responder Dispatching DSS will be deployed either within an emergency responder facility or a state or local government facility. Appropriate systems administrators, system maintenance, and IT personnel will be required. Suitable communications infrastructure will be required for the dissemination of alerts and advisories to emergency vehicle operators and emergency responder dispatchers.

Chapter 5 Operational Scenarios

This chapter develops operational scenarios for each high-priority application identified earlier in the document. The operational scenarios (also referred to as *use cases*) describe the ways in which the application will operate from the perspective of a typical user. Each operational scenario clearly describes the problem or situation that it is intended to address. Users are introduced, and their interactions with the system and the outcomes they experience are described.

Scenario for the Enhanced Maintenance Decision Support System

Description

In this scenario, connected vehicles, including participating winter maintenance vehicles, provide road weather information that is combined with other ESS, remote sensor, and meteorological model data form the inputs to the Enhanced-MDSS. Outputs from the Enhanced-MDSS are provided to state and local DOT winter maintenance personnel in the form of recommended treatment plans. In turn, the treatment plans are provided to the operators of snowplows and other winter maintenance vehicles. Figure 8 illustrates the scenario for the E-MDSS.

Chapter 5 Operational Scenatios

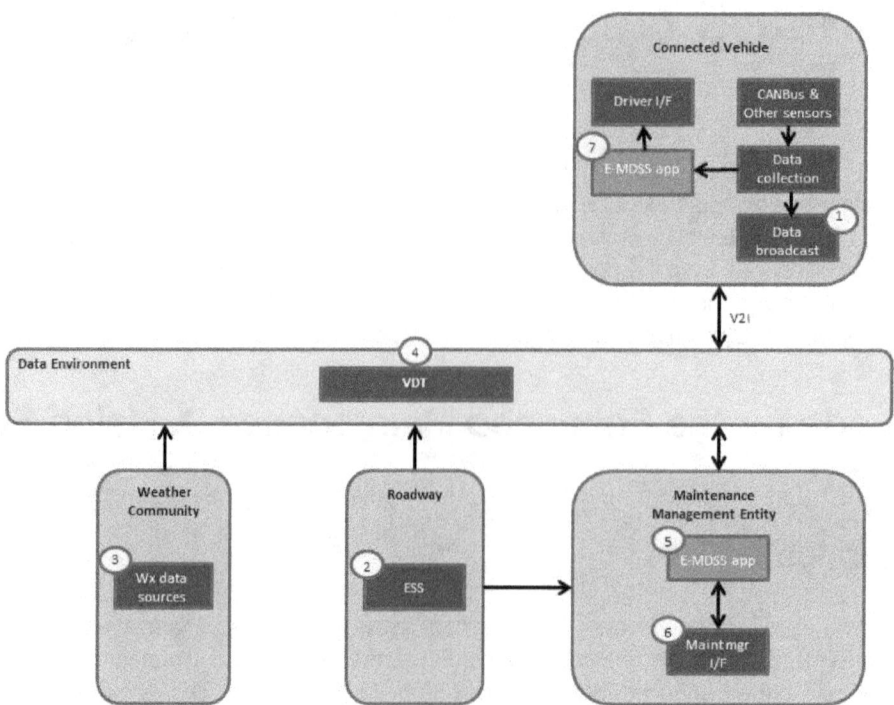

Figure 8: Scenario for the Enhanced Maintenance Decision Support System

Steps

1. Connected vehicles broadcast position and data that can be used to determine road and weather conditions via roadside equipment to the VDT within a data environment.

2. ESSs and other remote sensor systems send data to the VDT.

3. The VDT acquires forecast and other meteorological model output data.

4. The VDT ingests data, performs quality checks, applies algorithms, and outputs advanced road segment data to an Enhanced-MDSS application.

5. An Enhanced-MDSS application ingests advanced roadway segment data from the VDT, develops enhanced forecasts and road condition predictions, and determines recommended treatments plans.

6. An Enhanced-MDSS application outputs recommended treatment plans to a winter maintenance manager for approval.

7. Approved treatment plans are communicated to an in-vehicle Enhanced-MDSS application in snowplows and other winter maintenance vehicles for use by the vehicle operator.

Scenario for Information for Maintenance and Fleet-Management Systems

Description

In this scenario, connected maintenance vehicles of state and local DOTs provide data on location, asset and resource usage, material usage, and vehicle performance to Maintenance and Fleet-Management Systems. Output from these systems provides maintenance managers with the information needed for scheduling human resources, planning asset maintenance, budget planning and monitoring, life-cycle cost analyses, procurements, and purchasing decisions. Output from Maintenance and Fleet-Management Systems will also be used as input to the Enhanced-MDSS to identify the impacts of asset locations and the availability of winter treatment materials and chemicals on the recommended treatment plans. Figure 9 illustrates the Scenario for Information for Maintenance and Fleet-Management Systems.

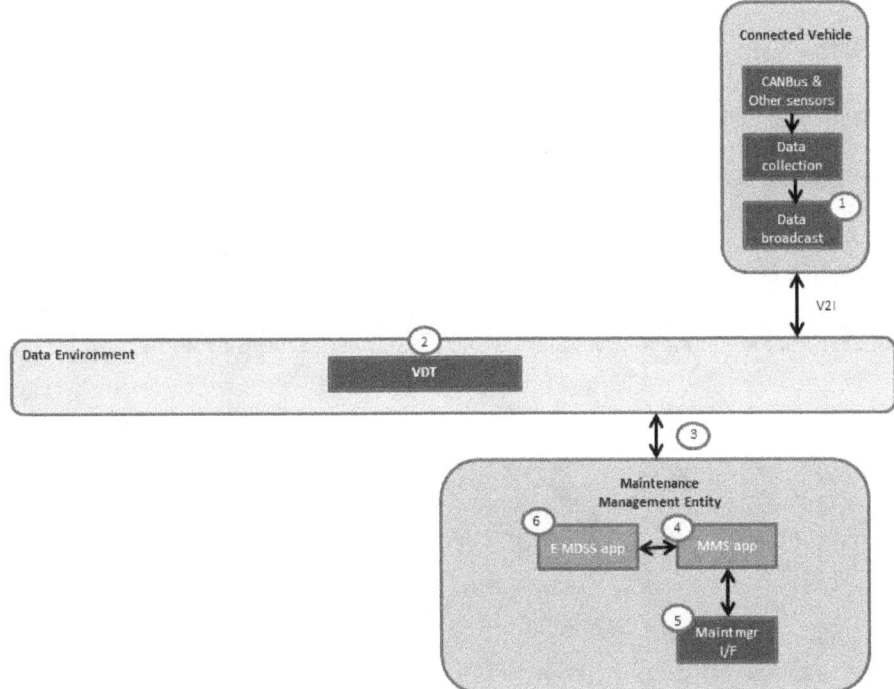

Figure 9: Scenario for Information for Maintenance and Fleet-Management Systems

Steps

1. Connected maintenance vehicles broadcast position, vehicle status, and material usage data via roadside equipment to a data environment.

2. Data are aggregated and organized in the data environment.

3. The Maintenance management entity receives the aggregated information from the data environment.

4. A Maintenance and Fleet-Management Application within the maintenance management entity uses the data in algorithms to produce maintenance and fleet-management metrics.

5. Maintenance and fleet-management metrics are presented to maintenance managers via interfaces in existing Maintenance and Fleet-Management Systems for use in planning, scheduling, and decision-making tasks.

6. The Enhanced-MDSS acquires metrics from the Maintenance and Fleet-Management application as inputs to the development of winter storm response plans and strategies.

Scenario for Variable Speed Limits for Weather-Responsive Traffic Management

Description

In this scenario, connected vehicles traveling on roadways upstream of a work zone provide location and road weather information to local and remote weather-responsive VSL applications. Locally processed data are displayed as a safe travel speed under the current road weather conditions on a roadside mobile speed trailer. With additional data inputs from ESSs, other remote sensors, and meteorological model outputs as well as additional processing through a weather-responsive VSL application at a remote location, refined speed advisories are sent to DMSs, work zone speed signs, and connected vehicle in-vehicle applications to advise motorists of safe speeds. Figure 10 illustrates the Scenario for Information for Maintenance and Fleet-Management Systems.

Chapter 5 Operational Scenatios

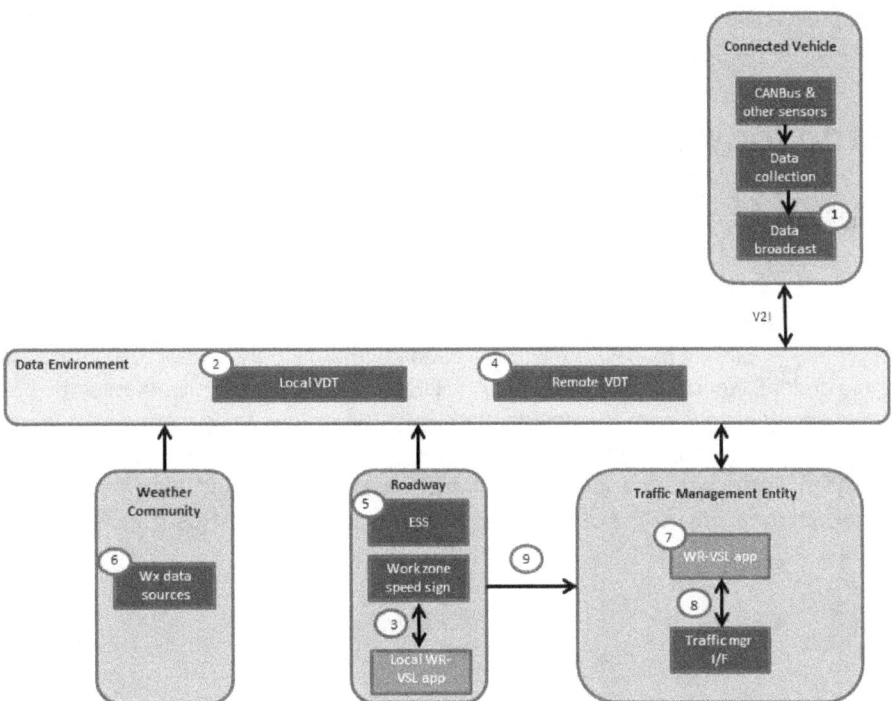

Figure 10: Scenario for Variable Speed Limits for Weather-Responsive Traffic Management

Steps

1. Connected vehicles broadcast position and data that can be used to determine road and weather conditions via roadside equipment to a local VDT processor at the roadside.

2. The VDT ingests data, performs quality checks, applies algorithms, and outputs near-real-time road segment data to a local weather-responsive VSL application.

3. The Local weather-responsive VSL application applies algorithms and outputs safe speed for display on a mobile work zone speed limit sign.

4. Connected vehicles broadcast position and data that can be used to determine road and weather conditions via roadside equipment to the remote VDT processor.

5. ESSs and other remote sensor systems send data to the VDT.

6. The VDT acquires forecast and other meteorological model output data.

7. The VDT ingests data, performs quality checks, applies algorithms, and outputs advanced road segment data to a weather-responsive VSL application in a traffic management entity.

8. The weather-responsive VSL application applies algorithms and outputs recommended safe speed to a traffic manager for approval.

9. Approved safe speed recommendations are communicated to mobile work zone speed limit signs, DMSs, and connected vehicle in-vehicle signs.

Scenario for Weather-Responsive Signalized Intersection

Description

In this scenario, connected vehicles traveling on roadways upstream of a suitably equipped signalized intersection provide location and road weather information to local and remote weather-responsive signal applications. Locally processed data are used to determine adjustments to intervals in the signal cycle under the current road weather conditions, which are then implemented by the traffic signal controller. With additional data inputs from ESSs, other remote sensors, and meteorological model outputs as well as additional processing through a weather-responsive signal application at a remote location, appropriate special signal timing plans are selected and implemented through the traffic signal controller. Figure 11 illustrates the scenario for a Weather-Responsive Signalized Intersection.

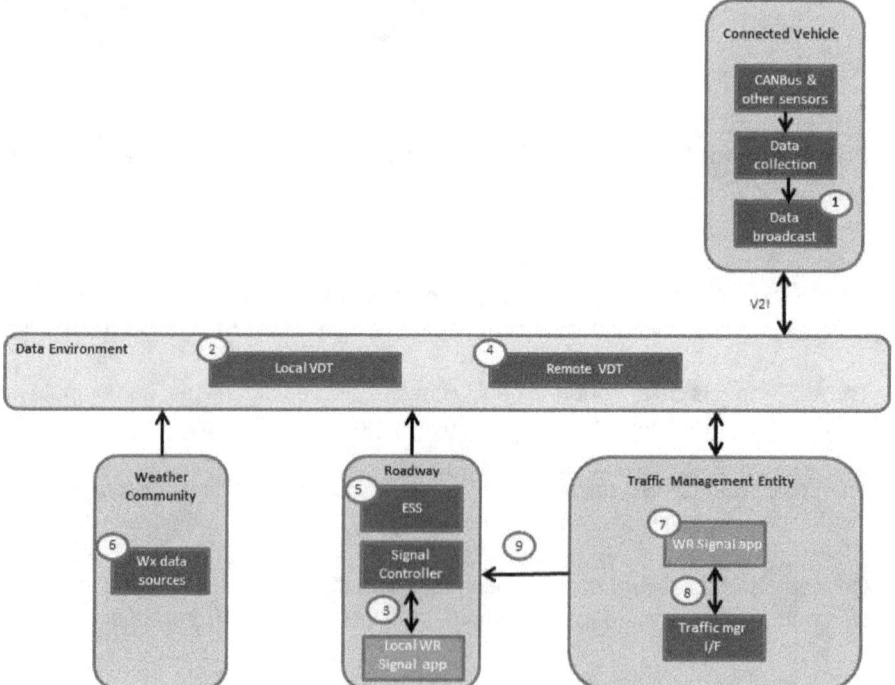

Figure 11: Scenario for Weather-Responsive Signalized Intersection

Steps

1. Connected vehicles broadcast position and data that can be used to determine road and weather conditions via roadside equipment to a local VDT processor at the roadside.

2. The VDT ingests data, performs quality checks, applies algorithms, and outputs near-real-time road segment data to a local weather-responsive signal application.

3. The local weather-responsive signal application applies algorithms and outputs adjusted signal intervals to the traffic controller.

4. Connected vehicles broadcast position and data that can be used to determine road and weather conditions via roadside equipment to the remote VDT processor.

5. ESSs and other remote sensor systems send data to the VDT.

6. The VDT acquires forecast and other meteorological model output data.

7. The VDT ingests data, performs quality checks, applies algorithms, and outputs advanced road segment data to a weather-responsive signal application in a traffic management entity.

8. The weather-responsive VSL application applies algorithms and outputs recommended special signal timing plan to a traffic manager for approval.

9. The approved signal timing plan is communicated to the traffic signal controller.

Scenario for Motorist Advisory and Warning System

Description

In this scenario, connected vehicles provide location and road weather information to a Motorist Advisory and Warning System application. The Motorist Advisory and Warning System application use algorithms to process the data and issue short time horizon advisories and warnings to participating connected vehicle drivers. The Motorist Advisory and Warning System application also assimilates the connected vehicle data with road condition and meteorological data from other sources to produce roadway segment-specific medium and long time horizon advisories and warnings that are communicated to participating users and connected vehicle drivers. Figure 12 illustrates the scenario for the Motorist Warning Advisory and Warning System.

Chapter 5 Operational Scenatios

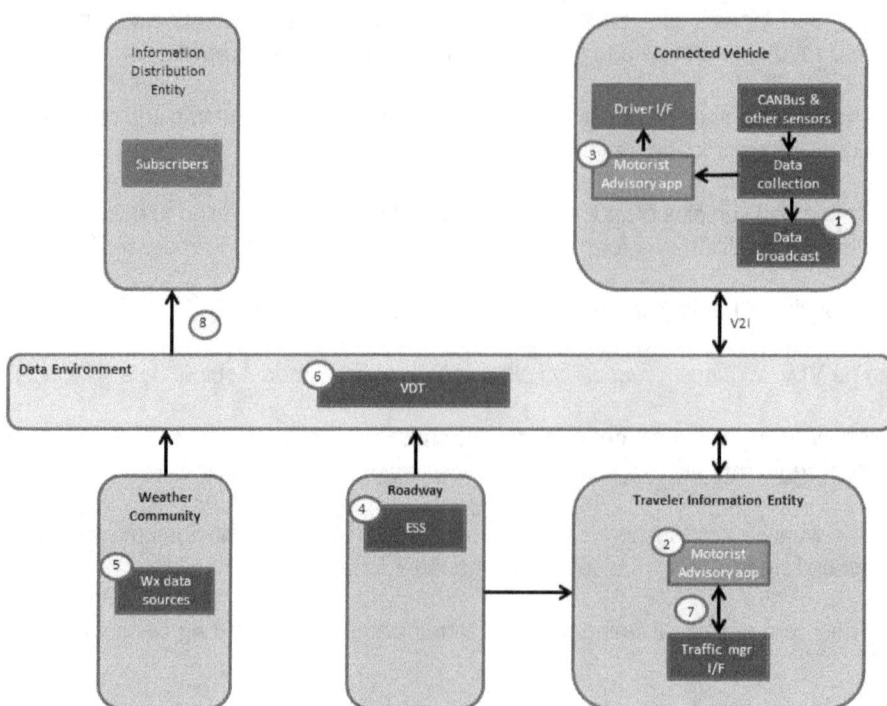

Figure 12: Scenario for Motorist Advisory and Warning System

Steps

1. Connected vehicles broadcast position and data that can be used to determine road and weather conditions via roadside equipment to a VDT processor.

2. The VDT ingests data, performs quality checks, applies algorithms, and outputs near-real-time road segment data to a Motorist Advisory and Warning System application.

3. The Motorist Advisory and Warning System application applies algorithms and outputs short time horizon advisories and warning to in-vehicle systems in participating connected vehicles.

4. ESSs and other remote sensor systems send data to the VDT.

5. The VDT acquires forecast and other meteorological model output data.

6. The VDT ingests additional data, performs quality checks, applies algorithms, and outputs advanced road segment data to the Motorist Advisory and Warning System application in a traffic management entity.

7. The Motorist Advisory and Warning System application applies algorithms and outputs medium and long time horizon advisories and warnings to a traffic manager for approval.

8. The approved medium and long time horizon advisories and warnings are communicated to participating entities for onward distribution to their users or subscribers.

Scenario for Information for Freight Carriers

Description

In this scenario, connected vehicles of all types provide location and road weather information to a Road Weather Advisory and Warning System for Freight Carriers application. The Road Weather Advisory and Warning System for Freight Carriers application use algorithms to process the data and issue short time horizon advisories and warnings specifically to participating truck drivers. The Road Weather Advisory and Warning System for Freight Carriers application also assimilates the connected vehicle data with road condition and meteorological data from other sources to produce roadway segment-specific medium and long time horizon advisories and warnings that are communicated to participating trucking firms or their service providers for integration with other routing, scheduling, or restriction information to create information that is relevant to the needs of motor carriers. Figure 13 illustrates the scenario for Information for Freight Carriers.

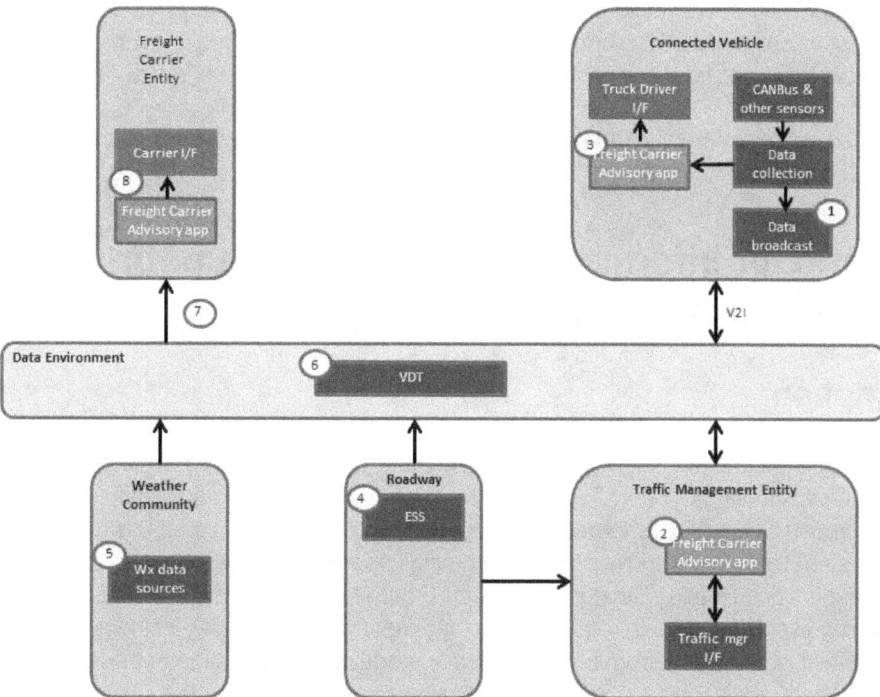

Figure 13: Scenario for Information for Freight Carriers

Steps

1. Connected vehicles of all types broadcast position and data that can be used to determine road and weather conditions via roadside equipment to a VDT processor.

2. The VDT ingests data, performs quality checks, applies algorithms, and outputs near-real-time road segment data to a Road Weather Advisory and Warning System for Freight Carriers application.

3. The Road Weather Advisory and Warning System for Freight Carriers application applies algorithms and outputs short time horizon advisories and warning to in-vehicle systems in participating trucks.

4. ESSs and other remote sensor systems send data to the VDT.

5. The VDT acquires forecast and other meteorological model output data.

6. The VDT ingests additional data, performs quality checks, applies algorithms, and outputs advanced road segment data to systems operated by motor carriers or their service providers for additional processing in a freight carrier entity.

7. Freight carrier applications assimilate additional data relating to schedule and routing restrictions, apply algorithms, and output medium and long time horizon advisories and warnings specific to the needs of participating carriers.

8. The approved medium and long time horizon advisories and warnings are communicated to participating motor carriers for distribution to their drivers.

Scenario for Information and Routing Support System for Emergency Responders

Description

In this scenario, connected vehicles of all types provide location and road weather information to an Information and Routing Support System for Emergency Responders application. The Information and Routing Support System for Emergency Responders application use algorithms to process the data and issue short time horizon advisories and warnings specifically to participating emergency vehicles. The Information and Routing Support System for Emergency Responders application also assimilates the connected vehicle data with road condition and meteorological data—including, in particular, wind and roadway flooding data—from other sources to produce roadway segment-specific medium and long time horizon advisories and warnings that are communicated to participating emergency service providers or their service providers for integration with other routing and scheduling information to create information that is relevant to the specific needs of emergency responders. Figure 14 illustrates the scenario for Information and Routing Support System for Emergency Responders.

Chapter 5 Operational Scenatios

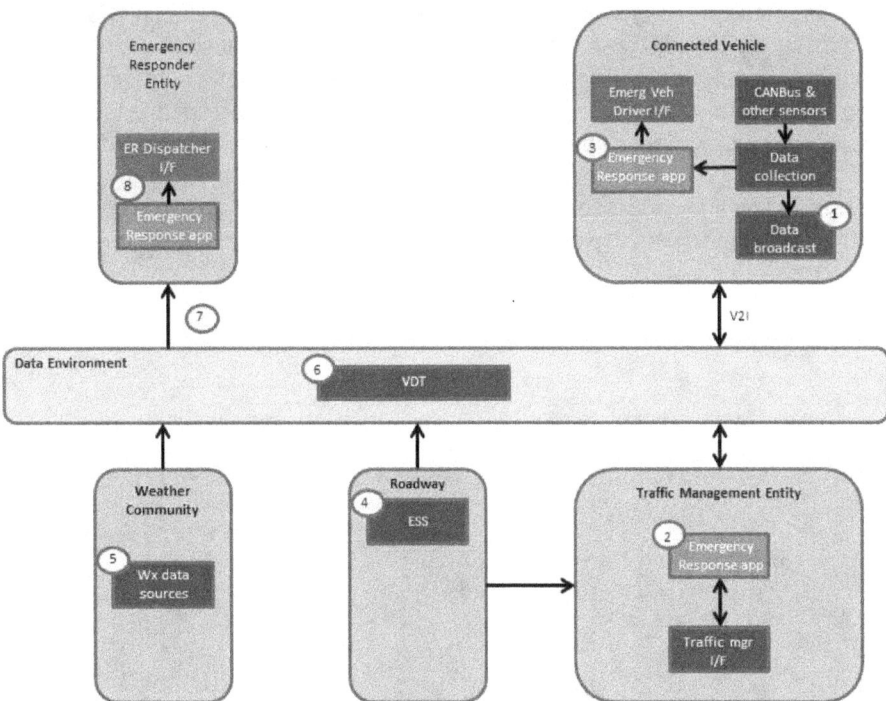

Figure 14: Scenario for Information and Routing Support System for Emergency Responders

Steps

1. Connected vehicles of all types broadcast position and data that can be used to determine road and weather conditions via roadside equipment to a VDT processor.

2. The VDT ingests data, performs quality checks, applies algorithms, and outputs near-real-time road segment data to an Information and Routing Support System for Emergency Responders application operated by an emergency responder or its service provider in an emergency response entity.

3. The Information and Routing Support System for Emergency Responders application applies algorithms and outputs short time horizon advisories and warning to in-vehicle systems in participating emergency vehicles.

4. ESSs and other remote sensor systems send data, including wind and roadway flooding data, to the VDT.

5. The VDT acquires forecast and other meteorological model output data.

6. The VDT ingests additional data, performs quality checks, applies algorithms, and outputs advanced road segment data to the Information and Routing Support System for Emergency Responders for additional processing in the emergency response entity.

7. The Information and Routing Support System for Emergency Responders application assimilates additional data relating to roadway restrictions caused by weather events, emergency vehicle routing, response times, and availability of emergency responders in neighboring jurisdictions; applies algorithms; and outputs medium and long time horizon advisories and warnings specific to the needs of the emergency responder.

8. Approved medium and long time horizon advisories and warnings are communicated to participating emergency response dispatchers for onward distribution to emergency vehicle drivers.

Chapter 6 Summary of Impacts

This chapter provides a summary of key impacts identified during the project.

Operational impacts

Among the six operational concepts identified and described in this document, several operational impacts have been identified:

- **Need for a Connected Vehicle Infrastructure.** Operational impacts will include the need to deploy, operate, and maintain the connected vehicle field infrastructure; the in-vehicle components in state and locally owned vehicles, including snowplows; and the associated data communications and processing systems. Motor carriers and emergency responders participating in the relevant operational concepts will be required to implement the appropriate in-vehicle components in their vehicles.

- **Implementation of New Systems.** Required new systems will include the VDT and the various systems for each of the identified applications. Implementation of new systems will affect state and local agencies, service providers for motorist advisories and warnings, freight carriers and their service provides, and emergency responders and their service providers.

- **Changes to Existing Systems.** The operational concepts indicate that changes would be required to the existing federal prototype MDSS, existing fleet and maintenance management systems, existing traffic-management systems for the VSL application, and the existing dispatching systems of freight carriers and emergency responders. It is assumed that further enhancements to the VDT would be required to accommodate the applications described in this document.

- **New Data Sources and Data Processing Capabilities.** Each operational concept relies on the availability of connected vehicle data. In addition, other sources of road weather and meteorological data are generally required by the applications. These sources of data may be greater than current data availability from ESSs. When the *Clarus* System is no longer available to state and local agencies after 2012, state and local agencies and other participants in these applications may need to seek a new mechanism for obtaining regional road weather data. MADIS may satisfy this need but will require further investigation.

- **New Operational Procedures.** The new capabilities that will emerge from proposed applications will demand the development of new operational procedures by state and local agencies, emergency responders, and potentially by freight carriers. In particular, the new applications suggest new maintenance procedures (including winter maintenance procedures); new procedures for asset management, budgeting, purchasing, and purchasing decisions within maintenance organizations; new traffic-management procedures relating to weather-responsive VSLs and motorist advisories and warnings; and new dispatching procedures for emergency responders and, potentially, freight carriers.

- **New Training Requirements.** New systems within the maintenance and traffic-management organizations of state and local agencies and emergency responder organizations infer the need for additional training in the use of the systems.

Organizational Impacts

Among the six operational concepts identified and described in this document, the following organizational impacts have been identified:

- **New Interactions Among Public Agencies.** The applications described in this document identify expanded interactions among divisions within state and local transportation agencies, particularly between maintenance management and traffic-management entities, to support weather-responsive VSLs and motorist advisories and warnings. Expanded or new interactions among agencies are also indicated, such as between transportation agencies and emergency responders.

- **New Interactions with Private Entities.** The applications identify several situations in which segment-specific road weather information is potentially provided to private-sector service providers in the areas of motorist advisories, freight information, and emergency responder information.

Chapter 7 Analysis of the Proposed System

This chapter provides an analysis of the benefits, limitations, advantages, and disadvantages of the proposed applications.

Summary of Improvements

The road weather connected vehicle applications will bring out transformative changes in the areas of safety, mobility, productivity, and reliability. Improvements that can be expected from the applications include—

- Improved driver awareness of road and weather conditions, leading to reduced crashes and fatalities

- Improved information about current and forecasted road and weather conditions, allowing drivers to better plan trips and make informed decisions about taking or deferring trips

- Improved ability by maintenance agencies to keep roadways clear and to improve road conditions during winter storms, leading to improved roadway safety and mobility for users

- Improved information availability for maintenance agencies to make better decisions about asset and resource availability and scheduling, equipment maintenance, and equipment and materials purchasing or procurement, leading to greater agency efficiency and productivity

- Improved ability to advise motorists on safe travel speeds during adverse weather conditions, especially in challenging situations such as work zones

- Improved ability to respond to the specific road weather information needs of motor carriers, with consequent safety and productivity gains for freight carriers

- Improved ability to respond to the specific road weather information needs of emergency responders, with consequent safety benefits to emergency vehicle drivers and better routing and response capabilities to emergency calls, with further potential to save lives.

Disadvantages and Limitations

This CONOPS has identified six road weather connected vehicle applications that are considered high priorities and merit further attention. Given the breadth of the applications, this document has been able to address each application at a high level. Additional definition of the applications will be required as the development process proceeds.

At this stage, two significant limitations have been identified that may affect further progress:

- **Need for Additional System Development.** Each application requires the development of new system capabilities, which will take time and investment. Additional development to add new capabilities to existing systems is also called for. In particular, the addition of new capabilities to the VDT is identified for each application, while enhancements to the existing federal prototype MDSS and to the various maintenance management systems agencies use will also be required. New data-processing capabilities and the necessary research and development of suitable algorithms to turn connected vehicle data and other road and weather data into meaningful and actionable information will be necessary. Suitable decision support tools for the various system users must also be developed.

- **Need for Close Interaction with New Stakeholder Communities.** Two of the identified applications will require a development effort that involves the deep participation of the freight carrier and emergency responder communities. Although the outreach efforts of the RWMP have started to engage these stakeholders, there will need to be a much expanded effort to more deeply understand the needs and capabilities of these groups from organizational, operational, and system availability standpoints.

References

The following references were used in the development of this document:

Mitretek Systems, "Weather Impacts on Arterial Traffic Flow," December 24, 2002.

Noblis, Inc., "Concept of Operations and Implementation Options for MDSS/MMS Data Exchange," September 2009.

U.S. Department of Transportation, Federal Highway Administration, "Clarus. A Clear Solution for Road Weather Information," Fact Sheet, 2011.

U.S. Department of Transportation, Federal Highway Administration, "Developments in Weather Responsive Traffic Management Strategies," Prepared by Battelle, Report No. FHWA-JPO-11-086, June 2011.

U.S. Department of Transportation, Federal Highway Administration, "Road Weather Information System Environmental Sensor Station Siting, Guidelines, Version 2.0," Prepared by SAIC, Report No. FHWA-HOP-05-026, November 2008.

U.S. Department of Transportation, Federal Highway Administration, "Vision and Operational Concept for Enabling Advanced Traveler Information Services," Revised Draft Report, Prepared by Kimley-Horn and Associates and Cambridge Systematics, April 4, 2012.

U.S. Department of Transportation, Federal Highway Administration and Missouri Department of Transportation, "Missouri Weather Response System Concept of Operations," Prepared by Mixon Hill, November 2004.

U.S. Department of Transportation, Research and Innovative Technologies Administration, "How Connected Vehicles Work," Fact Sheet, 2011.

U.S. Department of Transportation, Research and Innovative Technologies Administration, "Passenger Bus Industry Weather Information Application," Prepared by Global Science & Technology, Inc., Report No. FHWA-JPO-11-123, March 2011.

U.S. Department of Transportation, Research and Innovative Technologies Administration, "Concept Development and Needs Identification for Intelligent Network Flow Optimization (INFLO)," Prepared by SAIC, Draft Report, March 2012.

References

U.S. Department of Transportation, Research and Innovative Technologies Administration, "Concept of Operations for the Use of Connected Vehicle Data in Road Weather Applications," Prepared by University Corporation for Atmospheric Research, Draft Report, December 2011.

U.S. Department of Transportation, Research and Innovative Technologies Administration, "Freight Advanced Traveler Information System," Prepared by Cambridge Systematics, Final Report, March 2012.

U.S. Department of Transportation, Research and Innovative Technologies Administration, "The Vehicle Data Translator V3.0 System Description," Prepared by University Corporation for Atmospheric Research, Report No. FHWA-JPO-11-127, May 2011.

U.S. Department of Transportation, Research and Innovative Technologies Administration, "Vehicle Information Exchange Needs for Mobility Applications," Prepared by Noblis, Report No. FHWA-JPO-12-021, February 2012.

APPENDIX A. Acronyms

ABS	Antilock Braking System
ATDM	Active Traffic and Demand Management
BSM	Basic Safety Message
CANBus	Controller Access Network Bus
CONOPS	Concept of Operations
CVS	Connected Vehicle System
DMA	Dynamic Mobility Applications
DMS	Dynamic Message Sign
DOT	Department of Transportation
DSRC	Dedicated Short-Range Communications
DSS	Decision Support System
EMS	Emergency Medical Services
EnableATIS	Enable Advanced Traveler Information Systems
ESS	Environmental Sensor Station
ETA	Estimated Time of Arrival
FHWA	Federal Highway Administration
FRATIS	Freight Advanced Traveler Information System
HAR	Highway Advisory Radio
HAZMAT	Hazardous Material
INFLO	Intelligent Network Flow Optimization
MADIS	Meteorological Assimilation Data Ingest System
MDSS	Maintenance Decision Support System
MoPED	Mobile Platform Environmental Data
NCAR	National Center for Atmospheric Research
NOAA	National Oceanographic and Atmospheric Administration
NWS	National Weather Service
RWIS	Road Weather Information System
RWMP	Road Weather Management Program
SAE	Society of Automotive Engineers
TOC	Traffic Operations Center

Appendix A

UI	User Interface
USDOT	U.S. Department of Transportation
V2I	Vehicle-to-Infrastructure
V2V	Vehicle-to-Vehicle
VDT	Vehicle Data Translator
VSL	Variable Speed Limit
WRTM	Weather-Responsive Traffic Management

APPENDIX B. Data in BSM Part 1 or Part 2

System	Data Elements	BSM Part 1	BMS Part 2	Wx. Related
All Vehicles				
	Brake system status	X		
	Position (local 3D)	X		
	Vehicle size	X		
	Motion	X		
	Ambient air temperature		X	X
	Ambient air pressure		X	X
	ABS active over 100 msec		X	X
	Exterior lights (status)		X	X
	Lights changed		X	X
	Rain sensor		X	X
	Road coefficient of friction		X	X
	Traction control system active over 100 msec		X	X
	Wiper status		X	X
	Wipers changed		X	X
	Airbag deployment		X	
	Azimuth to obstacle on the road		X	
	Confidence position		X	
	Confidence speed/heading/throttle		X	
	Confidence time		X	

Appendix B

	Date/time of obstacle detection	X
	Distance to obstacle on road	X
	Hazard lights active	X
	Level of brake application	X
	Recent or current hard braking	X
	Stop line violation	X
	Throttle position (percent)	X
	Vehicle data	X
	Vehicle type (fleet)	X
	Crash delta V	
	Estimated point of impact	
	Number of occupants	
	Occupant medical data	
	Occupant safety belt use	
	Owner ID	
	Toll payment	
	Vehicle fuel type	
	Vehicle ID	
	Vehicle log, including time, location, and detection	
	Vehicle resting position	
Emergency Vehicles (only)		
	Light bar in use	X
	Public safety vehicle responding to	X

Appendix B

Freight Vehicles (only)	emergency	
	Siren in use	X
	Approach road to intersection	
	Intended turning movement at intersection	
	Descriptive vehicle identifier	X
	Fleet owner code	X
	Hazardous material (HAZMAT) status	X
	Trailer weight	X
	Vehicle height	X
	Vehicle mass	X
	Vehicle placarded as HAZMAT carrier	X
	Vehicle type	X
	Destination and stops	
	Electronic manifest	
	Load matching request	
	Pickup or drop-off time request	
Light Vehicles (only)		
	Cost	
	Departure location	
	Desired mode	
	Destination	
	Destination	

Appendix B

	Estimated time of arrival (ETA) at destination	
	ETA for pickup	
	Evacuation information request	
	Number of occupants in vehicle	
	Origin	
	Ride sharing response	
	Selected route and mode	
	Target arrival time	
	Target departure time	
Maint. Vehicles (only)		
	Maintenance activities	X
	Segment and lanes plowed	X
	Type and amount of road chemical applied	X
Transit Vehicles (only)		
	Connection protection request	
	Current itinerary	
	Passenger count	
	Status versus schedule	
	Transit service type	

APPENDIX C. J2735 SE and Weather Data

This table summarizes the responses to the USDOT request to incorporate weather data into J2735 SE. The responses align with the current (February 15, 2012) working draft version of the J2735 SE Design Document.

Sampling rate: The original USDOT request was for a sampling rate of 1 Hz. The sampling rate of the individual sensors within the vehicle is beyond the scope of the project. The intention behind the design is that the latest reading from the sensor is included in the snapshot of data. The rate at which snapshots are taken varies according to the vehicle movement and behavior. However, under default conditions, a snapshot will be taken at an interval that varies between 4 and 20 seconds.

Note: *NS* signifies that there is *No Sensor* in vehicles that measure this parameter, nor is it considered that there will be in the near future. Therefore, we do not see any reason to include this variable in J2735 SE.

	Original Request from USDOT				Response	
Variable	Name	Description	Valid Range	Resolution	Current Included in J2735 v 2	Comments
Atmospheric Pressure	AtmosphericPressure	The force per unit area exerted by the atmosphere in 1/10ths of millibars, a.k.a. tenths of hectoPascals. A value of 65535 shall indicate an error condition or missing value.	650.0–1200.0mb	INTEGER (0..65535)	Range 580–1090 Resolution 2 hPa	This information was developed in conjunction with Noblis (A. Stem) and NCAR
Spot Wind Direction	WindDirection	The direction from which the wind is blowing, measured in degrees clockwise from true north. A value of 361 shall indicate an error condition or missing value. The wind direction shall be corrected for vehicle movement.	0–359°	INTEGER (0..361)	NS	—
Spot Wind Speed	WindSpeed	The wind speed in tenths of meters per second. The value of 65535 shall indicate an error condition or missing value. The wind speed shall be corrected for vehicle movement.	0.0–250.0 m/s	INTEGER (0..65535)	NS	—

Appendix C

Variable	Original Request from USDOT				Response	
	Name	Description	Valid Range	Resolution	Current Included in J2735 v 2	Comments
Air Temperature	AmbientAirTemperature	The air temperature in tenths of degrees Celsius. The value 1001 shall indicate an error condition or missing value.	–100.0°C to 100.0°C	INTEGER (–1000..1001)	–40°C to 150°C resolution 1 degree	—
Dew Point Temperature	AmbientDewpointTemp	The dew point temperature in tenths of degrees Celsius. The value 1001 shall indicate an error condition or missing value.	–100.0°C to 100.0°C	INTEGER (–1000..1001)	NS	—
Solar Radiation	SolarRadiation	The ultraviolet, visible, and near-infrared (wavelength of <3.0 micrometers) radiation hitting the earth's surface in watts per square meter. The value of 701 shall indicate a missing value.	0–700 W/m²	INTEGER (0,701)	Arbitrary ranges from 0–7, with 0 = Complete Darkness, 1 = Minimal Sun Light, and 7 = Maximum Sun Light	Note that vehicle sensors are not quantitative and are used to turn on lights.
Total Radiation	TotalRadiation (replaces SunSensor)	The average total radiation hitting the earth's surface in watts per square meter. The value of 1001 shall indicate a missing value.	0–1000 W/m²	INTEGER (0,1001)	—	See above.
Visibility	Visibility	Surface visibility measured in tenths of a meter. The value 200001 shall indicate an error condition or missing value.	0.0–20000.0 m	INTEGER (0..200001)	NS	—
Surface Temperature	SurfaceTemperature	The current pavement surface temperature in tenths of degrees Celsius. The value 2001 shall indicate an error condition or missing value.	–100.0°C to 200.0°C	INTEGER (–1000..2001)	NS	—

Appendix C

	Original Request from USDOT				Response	
Variable	Name	Description	Valid Range	Resolution	Current Included in J2735 v 2	Comments
Precipitation Indicator	PrecipYesNo	Indicates whether the sensor detects moisture. *Precip* indicates that moisture is currently being detected; *noPrecip* indicates that moisture is not currently being detected; *error* means that the sensor is not connected, not reporting, or is indicating an error.	N/A	INTEGER {precip (1), noPrecip (2), error (3)}	Arbitrary ranges from 0–7, with 0 = no rain, 1 = Light mist, and 7 = Heavy Downpour	Note that this sensor is for rain and snow. Some sensors are resistors that are in contact with water. Others use changes of reflected light inside the windshield. Used for automatic wipers. Note that wiper status for both front and rear is included with a swipes-per-minute value.
Rainfall or Water Equivalent of Snow	PrecipRate	The rainfall, or water equivalent of snow, rate in tenths of grams per square meter per second. The value of 65535 shall indicate an error condition or missing value.	0.0–11.0 g/m²/s	INTEGER (0..65535)	NS	See above.

Appendix C

Variable	Original Request from USDOT					Response	
	Name	Description	Valid Range	Resolution	Current Included in J2735 v 2	Comments
Precipitation Situation	PrecipSituation (replaces RainSensor)	Describes the weather situation in terms of precipitation, with *Intensity* meaning— • Slight <2 mm/h water equivalent • Moderate ≥2 and <8 mm/h water equivalent • Heavy ≥8 mm/h water equivalent	N/A	INTEGER {other (1), unknown (2), noPrecipitation (3), unidentifiedSlight (4), unidentifiedModerate (5), unidentifiedHeavy (6), snowSlight (7), snowModerate (8), snowHeavy (9), rainSlight (10), rainModerate (11), rainHeavy (12), frozenPrecipitationSlight (13), frozenPrecipitationModerate (14), frozenPrecipitationHeavy (15)}	NS	—
Detected Friction	MobileFriction	Indicates the measured coefficient of friction in percent. The value 101 shall indicate an error condition or missing value.	0–100	INTEGER (0..101)	0 (frictionless to 0.98 resolution 0.2) Note that this may be NS in the near term, as currently it is not available data.	Note that the traction control, stability control, and ABS activations each create an event snapshot. These event snapshots are given the highest priority for transmission. In addition, they are given the highest priority for retention when snapshots are being removed.
Roadway Water Level Depth	RoadwayWaterLevel	Indicates the depth of the water on the roadway in centimeters. The value 256 indicates an error or missing value.	0–255 cm	BYTE (0..256)	NS	
Adjacent Snow Depth	AdjacentSnowDepth	The depth of snow in centimeters on representative areas other than the highway pavement, avoiding drifts and plowed areas. The value 256 indicates an error or missing value.	0–255 cm	BYTE (0..256)	NS	

Appendix C

Variable	Original Request from USDOT				Response	
	Name	Description	Valid Range	Resolution	Current Included in J2735 v 2	Comments
Roadway Snow Depth	RoadwaySnowDepth	The current depth of unpacked snow in centimeters on the driving surface	0–255 cm	BYTE (0..256)	NS	
Roadway Ice Thickness	RoadwayIceThickness	Indicates the thickness of the ice in millimeters. The value 256 shall indicate an error condition or missing value.	0–255 mm	BYTE (0..256)	NS